Sustainable Civil Infrastructures

Sustainable Infrastructure impacts our well-being and day-to-day lives. The infrastructures we are building today will shape our lives tomorrow. The complex and diverse nature of the impacts due to weather extremes on transportation and civil infrastructures can be seen in our roadways, bridges, and buildings. Extreme summer temperatures, droughts, flash floods, and rising numbers of freeze-thaw cycles pose challenges for civil infrastructure and can endanger public safety. We constantly hear how civil infrastructures need constant attention, preservation, and upgrading. Such improvements and developments would obviously benefit from our desired book series that provide sustainable engineering materials and designs. The economic impact is huge and much research has been conducted worldwide. The future holds many opportunities, not only for researchers in a given country, but also for the worldwide field engineers who apply and implement these technologies. We believe that no approach can succeed if it does not unite the efforts of various engineering disciplines from all over the world under one umbrella to offer a beacon of modern solutions to the global infrastructure. Experts from the various engineering disciplines around the globe will participate in this series, including: Geotechnical, Geological, Geoscience, Petroleum, Structural, Transportation, Bridge, Infrastructure, Energy, Architectural, Chemical and Materials, and other related Engineering disciplines.

More information about this series at http://www.springer.com/series/15140

John S. McCartney · Laureano R. Hoyos
Editors

Recent Advancements on Expansive Soils

Proceedings of the 2nd GeoMEast International Congress and Exhibition on Sustainable Civil Infrastructures, Egypt 2018 – The Official International Congress of the Soil-Structure Interaction Group in Egypt (SSIGE)

Springer

Editors
John S. McCartney
Department of Structural Engineering
University of California, San Diego
La Jolla, CA, USA

Laureano R. Hoyos
Department of Civil Engineering
University of Texas at Arlington
Arlington, TX, USA

ISSN 2366-3405 ISSN 2366-3413 (electronic)
Sustainable Civil Infrastructures
ISBN 978-3-030-01913-6 ISBN 978-3-030-01914-3 (eBook)
https://doi.org/10.1007/978-3-030-01914-3

Library of Congress Control Number: 2018957466

This Springer imprint is published by the registered company Springer Nature Switzerland AG
The registered company address is: Gewerbestrasse 11, 6330 Cham, Switzerland

Preface

This volume of technical papers includes research on an important topic in geotechnical engineering: the hydro-mechanical behavior of expansive soils. The twelve papers in this volume encompass experimental studies, numerical simulations, and field studies on this topic. This collection includes seven experimental studies focused on quantifying the stabilization of expansive soils using chemical admixtures and fibers. Three papers focus on new analyses to evaluate the deformation response of clays under different initial conditions. Finally, four studies include studies on the behavior of foundations on expansive soils, using both numerical and field monitoring techniques. The papers in this collection are representative of local challenges facing geotechnical engineers in the Middle East and Asia, but their contributions can also be extended to other regions of the world. The efforts of all authors and their peer reviewers, many of them from the Unsaturated Soils committee of the Geo-Institute of the American Society of Civil Engineers (ASCE), are gratefully acknowledged.

Contents

About the Editors

Biographical Sketch for John S. McCartney, Ph.D., P.E

a. Professional Preparation

University of Colorado Boulder, Boulder, CO, Civil Engineering, B.S. (high distinction), 2002
University of Colorado Boulder, Boulder, CO, Civil Engineering, M.S., 2002
University of Texas at Austin, Austin, TX, Civil Engineering, Ph.D., 2007

b. Appointments

07/18 – present, Professor and Department Chair, University of California San Diego (UCSD)
11/14 – 07/18, Associate Professor, University of California San Diego (UCSD)
07/13 – 08/14, Lyall Faculty Fellow, University of Colorado Boulder (CU)
05/13 – 10/14, Associate Professor, University of Colorado Boulder (CU)
08/08 – 05/13, Assistant Professor, University of Colorado Boulder (CU)
08/08 – 07/12, Barry Faculty Fellow, University of Colorado Boulder (CU)
08/07 – 08/08, Assistant Professor, University of Arkansas at Fayetteville (UA)

c. Publications

i. Five Works Most Related

1. Mun, W., Coccia, C.J.R. and McCartney, J.S. (2017). "Application of hysteretic trends in the preconsolidation stress of unsaturated soils." Geotechnical and Geological Engineering. Accepted. DOI: 10.1007/s10706-017-0316-7.
2. Coccia, C.J.R. and McCartney, J.S. (2016). "Thermal volume change of poorly draining soils I: Critical assessment of volume change mechanisms." Computers and Geotechnics. 80 (December), 26–40.
3. Coccia, C.J.R. and McCartney, J.S. (2016). "Thermal volume change of poorly draining soils II: Constitutive modelling." Computers and Geotechnics. 80(December), 16–25.
4. Alsherif, N.A. and McCartney, J.S. (2015). "Nonisothermal behavior of compacted silt at low degrees of saturation." Géotechnique. 65(9), 703–716.
5. Khosravi, A. and McCartney, J.S. (2012). "Impact of hydraulic hysteresis on the small-strain shear modulus of unsaturated soils." ASCE Journal of Geotechnical and Geoenvironmental Engineering. 138(11), 1326–1333.

ii. Five other Significant Works

1. Başer, T., Dong, Y., Moradi, A.M., Lu, N., Smits, K., Ge, S., Tartakovsky, D., and McCartney, J.S. (2018). "Role of water vapor diffusion and nonequilibrium phase change in geothermal energy storage systems in the vadose zone." Journal of Geotechnical and Geoenvironmental Engineering. 144(7), 04018038.
2. McCartney, J.S. and Murphy, K.D. (2017). "Investigation of potential dragdown/uplift effects on energy piles." Geomechanics for Energy and the Environment. 10(June), 21–28. DOI: 10.1016/j.gete.2017.03.001.
3. Ghaaowd, I., Takai, A., Katsumi, T., and McCartney, J.S. (2015). "Pore water pressure prediction for undrained heating of soils." Environmental Geotechnics. 4(2), 70–78.

4. Murphy, K.D., McCartney, J.S., and Henry, K.S. (2015). "Evaluation of thermo-mechanical and thermal behavior of full-scale energy foundations." Acta Geotechnica. 10(2), 179–195.
5. Stewart, M.A. and McCartney, J.S. (2014). "Centrifuge modeling of soil-structure interaction in energy foundations." ASCE Journal of Geotechnical and Geoenvironmental Engineering. 140(4), 04013044-1-11.

d. Synergistic Activities

1. Selected Honors and Awards:

- IGS Award, International Geosynthetics Society, 2018
- Outstanding Faculty Advisor Award, ASCE San Diego Chapter, 2017
- Walter L. Huber Civil Engineering Research Prize, ASCE, 2016
- Practical Journal Paper Award from TC106 on Unsaturated Soils, ISSMGE, 2014
- Arthur Casagrande Award, ASCE, 2013
- ASTM President's Leadership Award, ASTM, 2013
- Shamsher Prakash Found. Prize for Excellence in Teaching of Geotechnical Eng., 2012
- Young Professor Paper Award, Deep Foundations Institute, 2012
- J. James R. Croes Medal, ASCE, 2012
- NSF Faculty Early Development (CAREER) Award, 2011
- Best Paper of 2010 in Geosynthetics International, 2011
- Richard S. Ladd D18 Standards Development Award, ASTM, 2011
- International Geosynthetics Society (IGS) Young Member Award, 2008
- NSF Graduate Research Fellowship, 2002–2005

2. Service to Geo-Institute of the American Society of Civil Engineers (ASCE):

- Vice-Chair, Geo-Institute Unsaturated Soils Committee, 2013–present
- Member, Geo-Institute Geoenvironmental Committee, 2014–present

- Member, Geo-Institute Outreach and Engagement Committee, 2015–present
- Conference Co-Chair, PanAm Unsat 2017, Dallas, TX.
- Chair of sessions on thermal behavior of soils and geostructures at GeoCongress 2012, GeoCongress 2014, IFCEE 2015, GeoChicago 2016, Geotech. Front. 2017, IFCEE 2018
- Journal of Geotechnical and Geoenvironmental Eng., Associate Editor of the Year 2012

3. Service to International Geosynthetics Society (IGS):

- IGS North America President-Elect, 2017-present; Treasurer, 2015–2017
- Young Member Committee Co-Chair, GeoAmericas 2016, Miami, FL
- Technical Program Co-Chair, Geosynthetics 2015, Portland OR.

4. Service to ASTM International:

- Chair, ASTM D18.04 Subcom. on Hydrologic Properties and Hydraulic Barriers
- Past Chair, ASTM D18.09 Subcom. on Cyclic and Dynamic Properties of Soils
- Technical lead for 8 standards on hydraulic, thermal and dynamic properties of soil

5. Editorial Positions on International Journals:

- *Editor* (2015–Present), Journal of Geotechnical and Geoenvironmental Engineering, ASCE.
- *Editorial Board Member* (2011–Present), Geotechnical Testing Journal, ASTM.
- *Editorial Board Member* (2014–Present), Geosynthetics International, ICE.
- *Editorial Board Member* (2014–Present), Geomechanics for Energy and Env., Elsevier.
- *Editorial Board Member* (2014–Present), Computers and Geotechnics, Elsevier.
- *Editorial Board Member* (2016–Present), Canadian Geotechnical Journal, NRC.
- *Editorial Panel* (2016–2018), Géotechnique Letters, ICE
- *Editorial Board Member* (2014–2018), Soils and Foundations, Elsevier.

- *Associate Editor* (2016–2018), Geotechnical and Geological Engineering
- *Guest Editor* (2014), DFI Journal: "State of the Practice on Geothermal Foundations."
- *Guest Editor* (2014), Journal of Geotechnical and Geological Engineering: "Thermo-mechanical Response of Soils, Rocks, and Energy Geostructures."

Biographical Sketch for Laureano R. Hoyos, Ph.D., P.E., M.ASCE

Lead Campus Manager, Dwight D. Eisenhower Transportation Fellowship Program, FHWA
Professor, Department of Civil Engineering
The University of Texas at Arlington, Arlington, Texas 76019
Tel: (817) 272-3879; Fax: (817) 272-2630; Email: lhoyos@uta.edu

Professional Preparation:

Postdoctoral Research Fellow, Louisiana State University, 1999
Postdoctoral Research Fellow, Georgia Institute of Technology, 1999
Ph.D., Civil and Environmental Engineering, Georgia Institute of Technology, 1998
M.Sc., Civil and Environmental Engineering, Georgia Institute of Technology, 1996
M.Sc., Geotechnical Engineering, University of Puerto Rico, RUM, Mayaguez, PR, 1993
M.Sc., Highway Engineering, Universidad del Cauca, Popayán, Colombia, 1991
B.Sc., Civil Engineering, Universidad de la Costa, Barranquilla, Colombia, 1988

Appointments:

- Professor, Department of Civil Engineering, UTA, Texas, 2014–current
- Associate Professor, Department of Civil Engineering, UTA, Texas, 2005–2014

- Assistant Professor, Department of Civil Engineering, UTA, Texas, 2000–2005

Honors:

- Lockheed Martin Aeronautics Excellence in Teaching Award, College of Engineering, University of Texas at Arlington, 2014
- Research Excellence Award, Office of the Provost, University of Texas at Arlington, 2006, 2007, 2008, and 2009
- Outstanding Civil Engineering Instructor Award, Department of Civil Engineering, University of Texas at Arlington, 2005
- Outstanding Early Career Faculty Award, College of Engineering, University of Texas at Arlington, 2003

Postdoctoral/Student Supervision:

- Five Sponsored Visiting Research Scholars
- Nine Ph.D.s completed; four currently in progress
- Twenty M.S. theses completed; six currently in progress

Sample Publications

- Patil, U.D., Hoyos, L.R., and Puppala, A.J. (2016). "Characterization of compacted silty sand using a double-walled triaxial cell with fully automated Relative-Humidity control." Geotechnical Testing Journal, ASTM, 39(5), pp. 741–756.
- Patil, U.D., Hoyos, L.R., and Puppala, A.J. (2016). "Modeling essential elasto-plastic features of compacted silty sand via suction-controlled triaxial testing." International Journal of Geomechanics, ASCE, DOI: 10.1061/(ASCE) GM.1943-5622.0000726, pp. 1–22.
- Hoyos, L.R., DeJong, J.T., McCartney, J.S., Puppala, A.J., Reddy, K.R., and Zekkos, D. (2015). "Environmental geotechnics in the U.S. Region: A brief overview." Environmental Geotechnics, ICE Publishing, 2(EG6), pp. 319–325.

- Hoyos, L.R., Suescún-Florez, E.A., and Puppala, A.J. (2015). "Stiffness of intermediate unsaturated soil from simultaneous suction-controlled resonant column and bender element testing." Engineering Geology, Elsevier, 188(2015), pp. 10–28.
- Hoyos, L.R., Velosa, C.L., and Puppala, A. J. (2014). "Residual shear strength of unsaturated soils via suction-controlled ring shear testing." Engineering Geology, Elsevier, 172(2014), pp. 1–11.
- Hoyos, L.R., Pérez-Ruiz, D.D., and Puppala, A. J. (2012). "Modeling unsaturated soil response under suction-controlled true triaxial stress paths." International Journal of Geomechanics, ASCE, 12 (3), pp. 292–308.
- Hoyos, L.R., Pérez-Ruiz, D.D., and Puppala, A. J. (2012). "Refined true triaxial apparatus for testing unsaturated soils under suction-controlled stress paths." International Journal of Geomechanics, ASCE, 12(3), pp. 281–291.
- Hoyos, L.R., Velosa, C.L., and Puppala, A. J. (2011). "A servo/suction-controlled ring shear apparatus for unsaturated soils: Development, performance, and preliminary results." Geotechnical Testing Journal, ASTM, 34(5), pp. 413–423.
- Hoyos, L.R., Ordonez, C.A., and Puppala, A. J. (2011). "Characterization of cement-fiber-treated reclaimed asphalt pavement aggregates: Preliminary investigation." Journal of Materials in Civil Engineering, ASCE, 23(7), pp. 977–989.
- Yuan, D., Nazarian, S., Hoyos, L.R., and Puppala, A.J. (2011). "Evaluation and mix design of cement-treated base materials with high content of reclaimed asphalt pavement." Journal of the Transportation Research Board, TRR No. 2212, pp. 110–119.

Other Synergistic Activities

Lead Campus Manager, Dwight D. Eisenhower Transportation Fellowship Program, FHWA, 2016–2017

Associate Dean, Honors College, University of Texas at Arlington, 2008–2009

Member, Institutional Review Board, University of Texas at Arlington, 2006–2009
Faculty Advisor, Student Chapter, Society of Hispanic Professional Engineers (SHPE), University of Texas at Arlington, 2001–2009
Faculty Advisor, Student Chapter, Texas Society of Professional Engineers (TSPE), University of Texas at Arlington, 2007–2010
Chair, Graduate Studies Committee, Department of Civil Engineering, University of Texas at Arlington, 2015–current
Area Coordinator, Geotechnical Engineering Program, Department of Civil Engineering, University of Texas at Arlington, 2012–2014
Conference Chair, PanAm-UNSAT 2017: Second Pan-American Conference on Unsaturated Soils, the Geo-Institute of the ASCE, November 12–15, 2017, Dallas
Committee Chair, Unsaturated Soils, the Geo-Institute of the ASCE, 2013–current
Committee Member, AFS90, Chemical and Mechanical Stabilization of Soils, Transportation Research Board (TRB), National Academy of Sciences, 2002–2010
Committee Member, AFP60, Engineering Behavior of Unsaturated Soils, Transportation Research Board (TRB), National Academy of Sciences, 2001–2010
Associate Regional Editor, *Environmental Geotechnics*, Thomas Telford, Institution of Civil Engineers, London, U.K.
Editorial Board Member, *Geotechnical Testing Journal*, ASTM International, West Conshohocken, PA
Guest Editor, *International Journal of Geomechanics*, ASCE
Guest Editor, *Indian Geotechnical Journal*, Springer
Guest Editor, *Sustainable Civil Infrastructures*, Springer

Stabilization of Expansive Soil with Corex Slag and Lime for Road Subgrade

Radha J. Gonawala(✉), Rakesh Kumar, and Krupesh A. Chauhan

SV National Institute of Technology, Surat 395007, India
radhagonawala@gmail.com

Abstract. Huge quantities of industrial waste are generated every year all over the world. An enormous amount of this industrial waste material is getting discarded at the landfilling site. If examined properly, an appropriate amount of this material can be employed in road construction sector. Even in developing countries, only a specific kind of waste materials is underused and that too on an experimental basis. Present investigation describes the behavioral aspect of expansive soils mixed with industrial waste Corex slag with lime to improve the load-bearing capacity of the soil. Corex slag was mixed with the expansive soil in the range of 10% to 30% with an increment of 5%. Lime was added by 2% and 4% in the mix by the dry weight of soil. The physical and chemical properties of soil + Corex slag + lime mixes were determined. Various parameters like specific gravity, Atterberg limits, pH, compaction, UCS, and CBR were determined to understand the effect of these blends. The microstructural investigation of Soil, Corex slag, and the lime mix was done by using SEM and XRD technique. The specific gravity of combination was increased with increased Corex slag percentage. The MDD of soil + Corex slag + lime also increased as the Corex slag have high specific gravity. Admixing of all these stabilizers improves soaked CBR and UCS values. The addition of 25% Corex slag with 4% lime in the soil gives an optimum mix. The UCS value of the soil increased from 0.24 MPa to 1.09 MPa with the optimum combination. The CBR value of samples increased from 1.86% to 53.52% at the optimum mix proportion. The strength of expansive soil with the addition of Corex slag and lime had improved significantly due to the formation of CSH, and CASH gel as confirmed by XRD analysis. The increment in the strength explained by the changes in microstructure, observed from SEM analysis. This experimental analysis supports the usage of Corex slag for stabilizing expansive soil with some percentage of lime for subgrade layer in Highway construction.

1 Introduction

An enormous quantity of waste products, namely fly ash, steel slag, copper slag, are produced every year in the world from various industries. All these wastes are being dumped either near the plants or in lagoons that consume valuable land, which affect the environment (Sharma and Sivapullaiah 2016). The reduction in the availability of conventional materials leads researchers and government to suggest alternative materials for various construction works (Patel and Shahu 2015). Many works have been

J. S. McCartney and L. R. Hoyos (Eds.): GeoMEast 2018, SUCI, pp. 1–14, 2019.
https://doi.org/10.1007/978-3-030-01914-3_1

pursued to utilise these wastes as stabilizer such as fly ash (Sharma et al. 2012; Athanasopoulou and Sireesh et al. 2014; Mohammadinia et al. 2016); GGBS in concrete for replacing cement (Sekhar et al. 2017); steel slag in subbase construction (Kumar et al. 2006); and dolime as a stabilisation material (Patel and Shahu 2015). Significant work has been done in the past few years to save natural material by using these materials in various construction works. Bulk use of these industrial wastes can be achieved by using them as filling material or in road construction. Various methods are already available for waste management and disposable system. However, the rate of application of utilization of waste material in road construction is insignificant (Patel et al. 2012; Sharma and Sivapullaiah 2016 and Athulyu et al. 2017).

India is one among the most prominent generators of the industrial wastes among which fly ash and various types of slags are produced in vast quantity. Many efforts have been carried out by the road authorities to increase the application of industrial waste. However, the maximum usage of industrial waste in construction work is still minimal. The shortage of aggregates and rising in the cost of aggregates lead to the use of industrial wastes in the road construction (Shahu et al. 2013). Having a similar chemical composition of Ordinary Portland Cement (OPC), Corex slag is an industrial waste obtained by quenching molten iron by Corex method. However, from the steel industries utilization of slag is restricted to use of granulated blasted furnace slag up to 55% (Sharma and Sivapullaiah 2016).

This work is carried out to observe the effects of the Corex slag- lime stabilizing on the expansive soil. Along with the Index properties, compaction test, California Bearing Ratio (CBR) test, and unconfined compressive strength (UCS) tests were carried out on expansive soil + Corex slag +lime mixtures in different proportions. The influence of the adding of different percentages of lime on the UCS for soil- Corex slag mixture was also studied. Lime is used in the mix, as the industrial wastes need chemical activator to activate the cementation reactions. The X-ray diffraction (XRD) and scanning electron microscopy (SEM) studies were conducted to identify phase compositions and microstructural changes.

2 Background

The usage of industrial wastes is mainly subjected to their pozzolanic reactivity. The probable reason for their low utilization rate might be their slow reaction processes. Increasing the reactivity of waste materials would support their bulk use in geotechnical applications. The amount of the pozzolanic reaction in fly ash and different slags is due to free lime, silica and alumina present in them (Patel and Shahu 2016). Materials lacking with these compounds would not show any improvement in the physical properties. In these materials, the chemical activation is found by using additives like cement, lime, etc. as they are considered as a conventional stabilizer (Sharma et al. 2012). Due to the price escalation of chemical additives like lime and cement and depletion of natural resources, it is the time to reduce the usage of such materials (Patel and Shahu 2016). Studies have suggested that mixtures of different industrial wastes might be helpful to use in road construction (Kumar et al. 2006; Patel et al. 2012; Barisic et al. 2013). The stabilization of clayey soil using fly ash with lime showed an increase in UCS value up to 105.2 kPa and increase in CBR up to 5.7% by addition of 20% fly ash and 8.5% lime respectively (Sharma et al. 2012). The experimental study for hydraulically bound mixes for road construction using foundry sands and steel slags showed best mechanical results with 80% steel slag- 20% foundry sand having seven days UCS and Indirect Tensile Strength values up to 5.6 and 0.72 MPa respectively (Pasetto and Baldo 2015). Study on stabilizing lithomargic clay using GBFS and cement/lime described its usage for geotechnical applications. The UCS and Cohesion value gained up to 25% by GBFS replacement and at seven days curing CBR and angle of internal friction increased up to 50%. The study showed by adding 2 and 4% cement/lime could activate the pozzolanic reactions in the soil-GBFS mix (Sekhar et al. 2017). However, studies regarding changes in the microstructure and mineralogy did not mention.

3 Materials Characterization and Methodology

3.1 Raw Materials

Expansive clay was brought from Hazira, Surat, Gujarat, India. The soil dug out from a depth of about 1 m below to the ground level to avoid the organic matter. At the laboratory, the soil was dried, pulverized and then subjected to various experimental studies to find physical, chemical, geotechnical characteristics. The Corex slag used in research obtained from the Essar Steel Ltd. Hazira, which is in the Surat district of Gujarat state, India and confirms that material is environmentally harmless. The slag collected from the dump site of the plant. The Corex slag was dry and in sandy nature. Commercially available lime used in the study bought from the market having CaO more than 70% in it.

The specific gravity of Soil, Corex slag, and all mixtures was determined as per IS 2720 (Part-3) - 1980 in pycnometer. The liquid limit (LL) and plastic limit (PL) of Soil, and soil + Corex slag + lime mixture determined according to IS 2720 (Part-5), 1985. The liquid limit test was conducted by cone penetration method. pH test is performed on all the soil + Corex slag + lime mixes to determine the optimum requirement of stabilizer for soil. The physical properties of the materials used in the study are given in Table 1. The grain size analysis was done according to the IS: 2720 (Part-4)-1985 and the distributions of grain size for soil and Corex slag recognized by sieving through Indian Standard sieves. The grain size distribution curves of the soil and the Corex slag are in Fig. 1. From curves, it can be observed that the soil contains more fines of clay and silt-sized particles, whereas the Corex slag particles are of fine sand size.

The chemical compositions of Corex slag and lime are shown in Table 2. It shows that calcium oxide and silicon dioxide are low in Corex slag, whereas alumina oxide is high. This also shows that Corex slag has low calcium oxide content, whereas in lime calcium oxide is main content. However, when these two materials are combined, each will provide sufficient mineral to undergo pozzolanic reactions with soil.

Microstructural and mineralogical studies were conducted using SEM and XRD, respectively. The XRD patterns of soil and Corex slag are shown in Figs. 2 and 3, respectively. Quartz is the primary observable crystalline phases in the soil, as observed

Table 1. Physical properties of soil and Corex slag

Properties	Material	
	Soil	Corex slag
Colour	black	black
Specific gravity	2.55	2.91
Liquid limit (%)	62	—
Plastic Limit (%)	25	—
Plasticity Index (%)	37	—
IS Classification	CH	SW
FSI (%)	70	13
OMC (%)	17.9	—
MDD (gm/cc)	1.69	—
UCS (kg/cm^2)	2.4	—
pH	8.09	9.6
CBR (%)	1.86	—

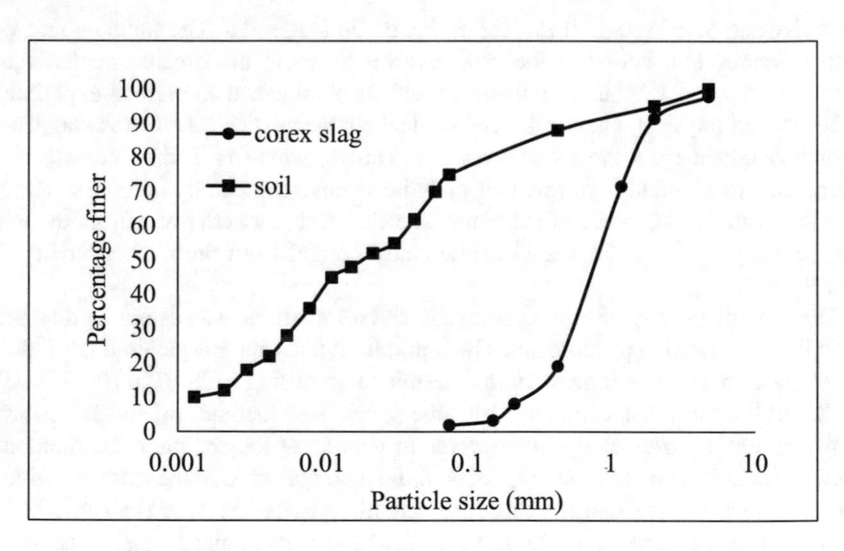

Fig. 1. Grain size distribution curves of Corex slag and soil

in Fig. 2 (Sharma and Sivapullaiah 2016). The peaks in soil were recognized, but not in the case of the Corex slag XRD patterns (Fig. 3), which showed considerable quantities of amorphous material. However, very few low-intensity peaks were identified.

3.2 Methodology

The experimental laboratory studies included the description of the materials and the assessment of their engineering properties, including compaction, CBR the UCS of the soil + Corex slag + lime combinations with varying percentages of Corex slag (10, 15,

Table 2. Chemical composition of Corex slag

Composition	Corex slag
Silica; SiO_2 (%)	3.91
Ferric Oxide; Fe_2O_3 (%)	3.15
Alumina; Al_2O_3 (%)	55.31
Calcium Oxide; CaO (%)	32.69
Magnesia; MgO (%)	0.74
Sulphuric Anhydride; SO_3 (%)	1.01
Insoluble Residue (%)	1.06
Loss on Ignition (%)	3.4

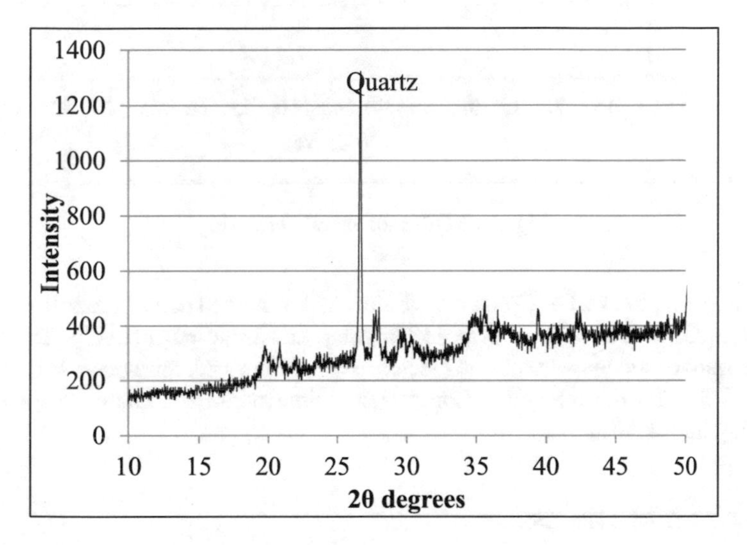

Fig. 2. XRD patterns of soil

20, 25 and 30%) and lime (2 and 4%). The portion of Corex slag added based on the total dry weight of the mix. For the compaction studies, the modified compaction test was used as per IS 2720 (Part-8)-1983. The UCS testing was carried out as per IS: 4332 (Part-5) - 1970. The UCS sample was prepared at their optimum moisture content (OMC) and maximum dry density (MDD). The specimens were 50 mm in diameter and 100 mm long. The samples were cured for 0, 7, 14 and 28 days. The prepared samples were wrapped in plastic and then kept in desiccators with 100% humidity. For all mixture of soil + Corex slag + lime, three similar specimens were tested and average UCS value determined. The effect of different percentage of Corex slag and lime on UCS also studied. The CBR test conducted on soil + Corex slag + lime mixes and three specimens were prepared for each combination by compacting at OMC and MDD. After compaction, the CBR specimens wrapped with polythene to maintain OMC and kept for curing for 3 days and tested after 4 days soaking.

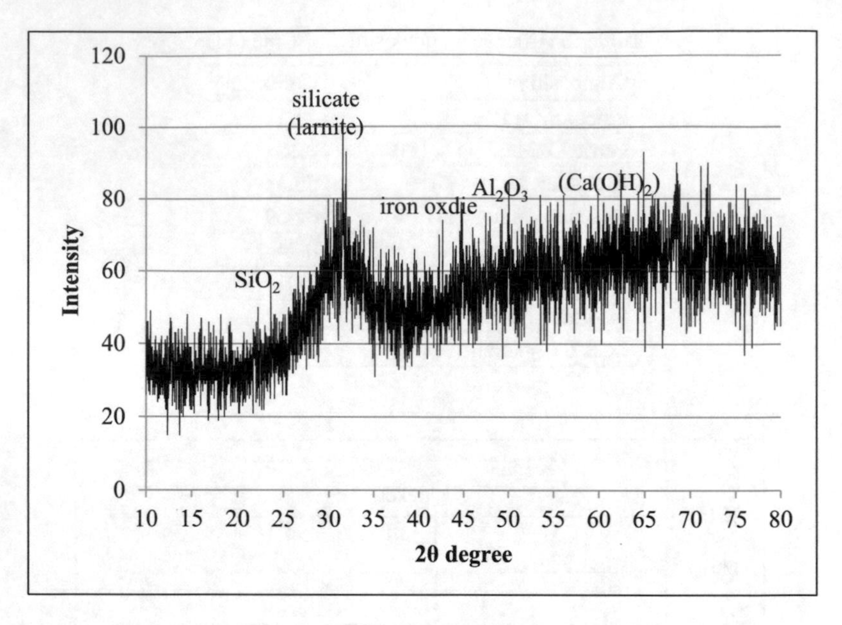

Fig. 3. XRD patterns of Corex slag

In this paper, the soil + Corex slag + lime mixes named as per the following label. Symbols S, CS and L refer to Soil, Corex Slag, and Lime respectively. The numbers prefix symbol specifies their respective weight in percentages. For example, 78S + 20CS + 2L means Soil + Corex slag + lime mixture contain 78% Soil, 20% Corex slag and 2% Lime.

4 Results and Discussion

4.1 Atterberg Limits

The changing in Atterberg limits with Corex slag and lime content tabulated in Table 3. Within rise of Corex and lime content, the plasticity index decreases with a decrease in liquid limit and increase in the plastic limit. Patel and Shahu, 2015 had, reported the same behavior with expansive soil stabilization with lime.

4.2 Free Swell Index (FSI)

FSI of the expansive soil decreases with increase in slag and lime content (Table 3). Soil, Slag, and lime particles coagulate and flocculate, which reduces the surface area of clay minerals leads to decreasing the quantity of water which can be absorbed by the clay mineral surfaces.

Table 3. Index properties of the different mixture

Combination	LL (%)	PL (%)	PI (%)	pH	Specific gravity	FSI (%)
88S + 10CS + 2L	42.4	29.89	12.51	11.51	2.58	43.00
83S + 15CS + 2L	40.2	28.5	11.7	11.68	2.62	38.00
78S + 20CS + 2L	35	25.5	9.5	11.81	2.65	21.50
73S + 25CS + 2L	31.5	22.12	9.38	11.92	2.66	19.45
68S + 30CS + 2L	29.45	22.97	6.48	12.25	2.68	29.30
86S + 10CS + 4L	39.3	28.52	10.78	11.98	2.6	17.50
81S + 15CS + 4L	35	28.32	6.68	12.35	2.61	14.50
76S + 20CS + 4L	33.2	27	6.2	12.44	2.63	11.12
71S + 25CS + 4L	31.9	26.43	5.47	12.59	2.65	11.97
66S + 30CS + 4L	27	22	5	12.66	2.66	17.52

4.3 Specific Gravity

Variation of the specific gravity of the soil + Corex slag + lime samples is presented in Table 3. Higher Corex slag content raised the specific gravity of the mixture as expected. It is due to the chemical composition of the mix, which is higher in oxide as Corex slag content increase.

4.4 pH Test

Change in pH was found due to cation exchange that took place in between soil, Corex slag, and lime, as clay minerals are linked with each other. The pH value of different mixtures is tabulated in Table 3. As per the IRC: 51-1992 stabilized clay should have minimum 12.4 pH value to consider as an optimum mix. Therefore from pH value, the combination of 71S + 25CS + 4L found as an optimum mix.

4.5 Compaction Characteristics

The OMC and MDD values of the mixtures at different percentages of Corex slag and lime are shown in Table 4. It was observed that MDD increases with an increase in the Corex slag content, where the variation in OMC with Corex slag content decreases. The higher specific gravity of Corex slag assists to increase in the MDD of the mixture rather than an increase in the degree of compaction (Patel and Shahu 2015).

4.6 UCS Test

The UCS test results of the soil + Corex slag + lime mixtures after different curing period shown in Figs. 4 and 5. The strength of blends increases between 0 to 7 days. A significant increase in strength was found after 28 days of curing for all combinations. It was noted that strength of the soil is getting increased as the percentage of Corex slag increases. In the presence of lime and silica, the pozzolanic reaction took place. In the combination of soil + Corex slag + lime, as the percentage of Corex slag and lime increases, both the materials provided sufficient calcium ions and alumina for

Table 4. OMC and MDD value of soil + Corex slag + lime mixture

Combination	OMC (%)	MDD (gm/cc)
88S + 10CS + 2L	18.84	1.88
83S + 15CS + 2L	16.88	1.89
78S + 20CS + 2L	16.57	1.90
73S + 25CS + 2L	15.40	1.91
68S + 30CS + 2L	14.20	1.93
86S + 10CS + 4L	16.27	1.82
81S + 15CS + 4L	16.00	1.86
76S + 20CS + 4L	15.10	1.89
71S + 25CS + 4L	13.50	1.92
66S + 30CS + 4L	12.50	1.95

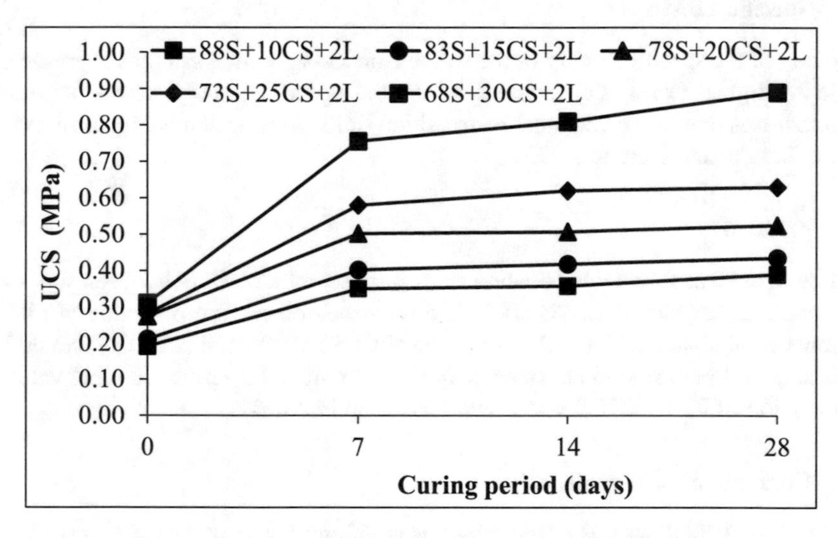

Fig. 4. Variation of UCS with curing period with 2% lime

the reaction with silica and got comparatively good strength (Patel and Shahu 2015). In pozzolanic reaction, the formation of cementitious compounds like calcium silicate hydrate (CSH) is a long-term process, and it depends on the curing period (Bastos et al. 2016).

Apart from the high alumina oxide content, Corex slag also contains significant calcium and silica oxide. So, an activator is required as the cementitious reactions are slow with Corex slag. The addition of lime helps to achieve strength within 7 days of curing period. The combining of 2% and 4% of lime increased the strength of all soil + Corex slag + lime mixtures. With lime, the hydration reaction was boosted, to form cementitious compounds such as CSH. The CSH gel provides early strength at early stages. The CSH gel is the reason for binding the particles and increasing the

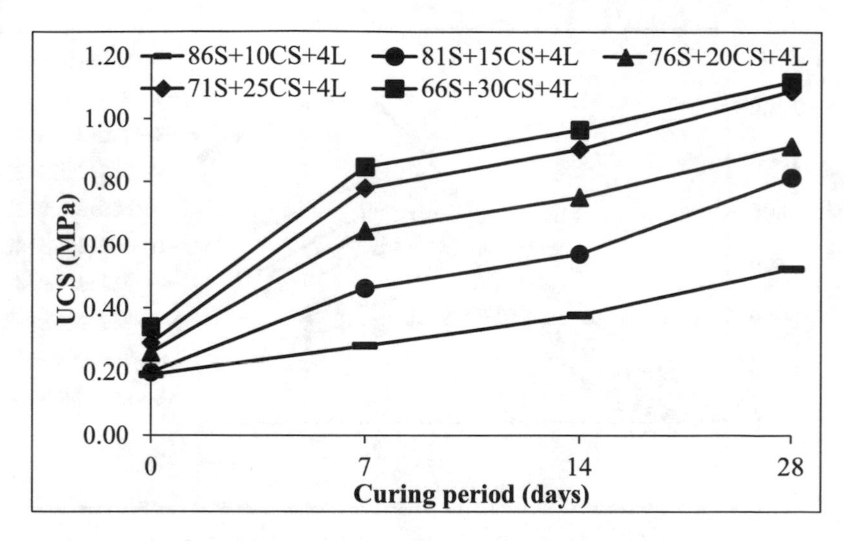

Fig. 5. Variation of UCS with curing period with 4% lime

strength of the mixtures (Bastos et al. 2016). The amount of gel formation increased as lime content increases and therefore the particles bound more efficiently with 4% lime content.

4.7 Stress-Strain Behavior

The stress-strain behavior of the samples of soil + Corex slag + lime is shown in Fig. 6. The samples get failure at a lower strain of brittle splitting type. After reaching the peak, samples showed quick drop-off to resist the load. The same stress-strain behavior showed by all combination of soil + Corex slag + lime mixture.

4.8 CBR Test

The CBR results for soil + Corex slag + lime stabilization support the increment strength capacity with the increase in binder content which is shown in Table 5. The strength increases with time due to presence lime; which reacts with the silica and alumina found in the Corex slag and soil, which develop significant cementing reactions and suggest pozzolanic activity. Results revealed that the addition of 10% slag with 4% lime corresponded to a 44.84% CBR value and further it tends to near 13% effective CBR. A per Indian specifications for flexible pavement design effective CBR should be more that 8% (IRC 37-2012). Therefore, the use of Corex slag in a subgrade layer material is suggested for road construction.

4.9 Microstructural Investigations

The soil + Corex slag +lime mixes were investigated by XRD tests and SEM. Diffraction peaks in the range of 10–60° (2θ) on the XRD spectrum were analyzed.

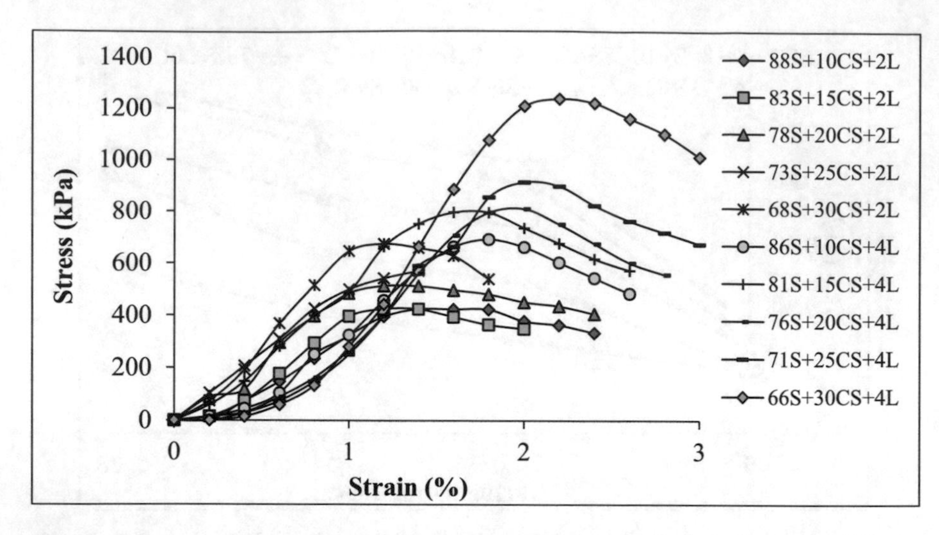

Fig. 6. Influence of Corex slag and lime on the axial stress-strain behavior from UCS test

Table 5. CBR value of soil + Corex slag + lime mixture

Combination	CBR (%)
88S + 10CS + 2L	17.87
83S + 15CS + 2L	22.21
78S + 20CS + 2L	21.58
73S + 25CS + 2L	38.04
68S + 30CS + 2L	34.32
86S + 10CS + 4L	44.84
81S + 15CS + 4L	48.38
76S + 20CS + 4L	52.43
71S + 25CS + 4L	53.52
66S + 30CS + 4L	66.14

The cured cylindrical specimens were broken from the Centre into small pieces, for SEM study.

4.10 XRD Analysis

The XRD pattern of 71S + 25CS + 4L for 28 days curing is shown in Fig. 7. The hydration products of Soil + Corex slag + lime include ettringite ($3CaO\ Al_2O_3\ 3CaSO_4\ 32H_2O$), CSH and calcium alumina silicate hydrate CASH. It was noticed that early strength development in the mix due to the formation of ettringite, whereas the later strength development of the blend occurs owing to the generation of CSH and CASH gel and same confirms in the other studies also (Bastos et al. 2016). Higher intensity peaks of the XRD pattern indicate higher percentages of corresponding compounds.

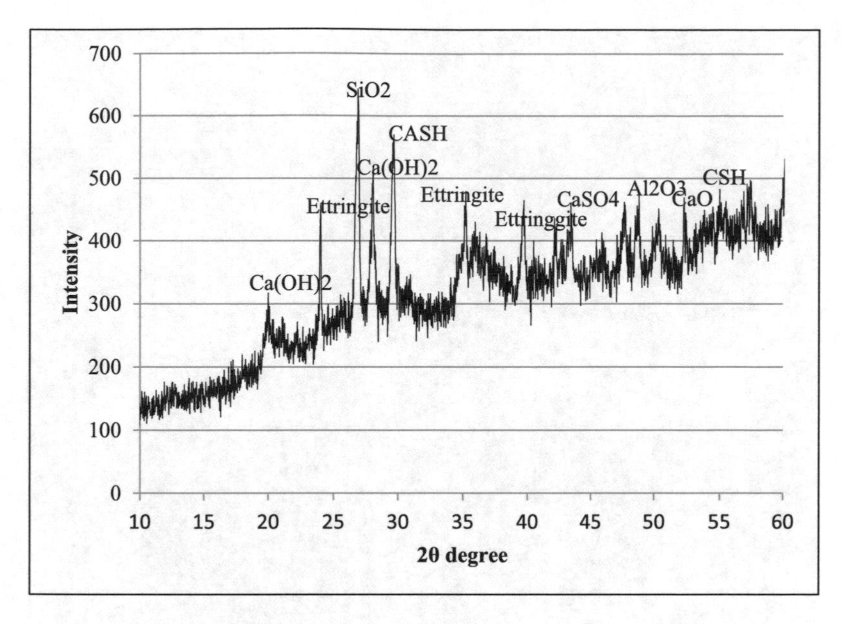

Fig. 7. XRD pattern of 71S + 25CS + 4L for 28 days

4.11 SEM Analysis

Figure 8 shows the SEM micrograph of 7 days cured samples of 71S + 25CS + 4L. The 7 day cured samples do not show a dense matrix, and particles can be identified. The formations of cementitious compounds are not high. It can enhance by curing the sample, which makes the samples denser, as seen from the micrograph Fig. 9. Stabilised samples show a compact interlinked structure because of the formation of cementitious compounds. The formation of higher amounts of hydration products corresponds to the higher compressive strengths. The cementitious mixtures start to develop in the samples, which binds the particles and improves strength.

5 Conclusions

The research work on the use of industrial waste Corex slag to stabilize subgrade soil with lime carried out in the laboratory and discussed. The conclusions drawn from the study are:

- The combinations of Soil + Corex slag + lime mixes gave a promising performance to act as a subgrade of flexible pavement in consideration of index properties.
- For soil + Corex slag + lime combinations, OMC decreases with increase in MDD. Soils stabilized with Corex slag, and lime increases the MDD and improves the workability of wet soil. As the quantity of Corex slag in the mixture increase, the effect of slag became more dominant in compare to lime. The dominance of Corex slag in blend tended to increase the MDD and decrease the OMC.

Fig. 8. SEM image of 71S + 25CS + 4L of 7 days in 10.00 μm

Fig. 9. SEM image of 71S + 25CS + 4L of 28 days in 10.00 μm

- The UCS and CBR results indicate that soil + Corex slag + lime has the potential to be used in highway subgrade construction using low amounts of lime. The UCS and CBR value increase with Corex slag content increase in the mixture. The UCS values increase rapidly with the increase in curing period as well.
- The addition of small amounts of lime increases the strength significantly. It is because the materials require initial activation to undergo the pozzolanic and hydraulic reactions.
- The effect of the addition of Corex slag, lime and the curing period of the mixture can be identified through microstructural analysis. The increase in the strength of the mix has been explained by the microstructural changes seen through SEM studies. Also, cementitious compounds, namely CSH, CASH in the stabilized samples, bind the particles, resulting in the stable microstructure.
- This experimental analysis could support the usages of waste material like Corex slag mixing with expansive soil for improving subgrade layer required strength by adding lime in Highway construction work.

References

Athanasopoulou, A.: Addition of lime and fly ash to improve highway subgrade soils. J. Mater. Civil Eng. **26**(4), 773–775 (2014). https://doi.org/10.1061/(ASCE)MT.1943-5533.0000856

Mohammadinia, A., Arulrajah, A., Sanjayan, J., Disfani, M.M., Bo, M.W., Darmawan, S.: Stabilization of demolition materials for pavement base/subbase applications using fly ash and slag geopolymers: laboratory investigation. J. Mater. Civil Eng. **28**(7), 1–9 (2016) https://doi.org/10.1061/(ASCE)MT.1943-5533.0001526. 04016033, ASCE

Sharma, A.K., Sivapullaiah, P.V.: Strength development in fly ash and slag mixtures with lime. In: Proceedings of the Institution of Civil Engineers Ground Improvement, vol. 169(GI3), pp. 194–205 (2016). http://dx.doi.org/10.1680/jgrim.14.00024

Sekhar, D.C., Nayak, S., Preetham, H.K.: Influence of granulated blast furnace slag and cement on the strength properties of lithomargic clay. Indian Geotech. J. **47**(3), 384–392 (2017). https://doi.org/10.1007/s40098-017-0228-8

Indian Standard designation IS 2720 Part III/Sec1: The method of test for soil, Determination of Specific Gravity, Section 1 Fine garined soils (1980)

Indian Standard designation IS 2720 Part IV: The method of test for soil, Grain size analysis (1985)

Indian Standard designation IS 2720 Part V: The method of test for soil, Laboratory Determination of Liquid & Plastic limits (1985)

Indian Standard designation IS 2720 Part VIII: The method of test for soil, Laboratory Determination of Moisture content & dry density (1985)

Indian Standard designation IS 2720 Part XVI. Method of test for soil, Laboratory determination of CBR (1985)

Indian Standard designation IS 4332 Part V: The method of test for stabilized soil, Determination of unconfined compressive strength of stabilized soils (1970)

IRC: 37-2012. Guidelines for the Design of Flexible Pavements (Third Revision) New Delhi, India

Shahu, J.T., Patel, S., Senapati, A.: Engineering properties of copper slag–fly ash–dolime mix and its utilization in the base course of flexible pavements. J. Mater. Civil Eng. **25**(12), 1871–1879 (2013). https://doi.org/10.1061/(ASCE)MT.1943-5533.0000756

Bastos, L.A.C., Silva, G.C., Mendes, J.C., Peixoto, R.A.F.: Using iron ore tailings from tailing dams as road material. J. Mater. Civil Eng. (2016). https://doi.org/10.1061/(ASCE)MT.1943-5533.0001613

Sharma, N.K., Swain, S.K., Sahoo, U.C.: Stabilization of a clayey soil with fly ash and lime: a micro level investigation. Geotech. Geol. Eng. **30**, 1197–1205 (2012)

Kumar, P., Chandra, S., Vishal, R.: Comparative study of different subbase materials. J. Mater. Civil Eng. **18**(4), 576 (2006). https://doi.org/10.1061/(ASCE)0899-1561

Patel, S., Shahu, J.T.: Engineering properties of black cotton soil-dolime mix for its use as subbase material in pavements. Int. J. Geomate. **8**(1), 1159–1166 (2015). (Sl. No. 15)

Effect of Sulfate Contamination on Compaction and Strength Behavior of Lime Treated Expansive Soil

P. Sriram Karthick Raja and T. Thyagaraj(✉)

Department of Civil Engineering, Indian Institute of Technology Madras, Chennai 600036, India
ttraj@iitm.ac.in

Abstract. Lime is the most commonly used chemical admixture for the treatment of the expansive soils. But the intrusion of sulfate contaminant into the lime treated soil will always results in deterioration of the treated soil. The intrusion of sulfate occurs through acid rains, effluent from tannery industries or mine wastes, intrusion of sea water, construction waste, intrusion of leachate from solid waste fills and sulphate rich groundwater. Therefore, this paper aims at bringing out the effect of sodium sulfate (Na_2SO_4) solution on the compaction and strength behavior of the lime treated expansive soil. Further, the effect of compaction delay and curing period on the sulfate contaminated lime treated soil is also brought out in this paper. Lime contents of 2.5 (initial consumption of lime (ICL) −1%), 3.5 (ICL) and 4.5% (ICL +1%) were used in this study. The sulfate contamination was limited to 5000, 10000 and 20000 ppm. The experimental results showed that the sulfate contamination decreased the maximum dry density and optimum moisture content of the lime treated expansive soil. Further, the intrusion of sulfate solution into the lime treated expansive soil decreased the strength of the lime treated expansive soil.

1 Introduction

Stabilization of expansive soil is vital in applications such as pavements, runways, earthen dams and erosion control (Diamond 1975; Yadima and Nwaiwu 2001; Osinubi and Nwaiwu 2006). Stabilization of expansive soil is brought about by mechanical or chemical stabilization methods, by which the engineering properties of the natural soil are improved. The mechanical stabilization involves the reinforcement ofsoil with geosynthetics, cohesive non-swelling soil cushion, preloading, etc., whereas the chemical stabilization involves the addition of lime, cement, bitumen, calcium chloride, polymers, etc. Among all the above methods for soil stabilization, lime stabilization is the most widely used method of stabilization by virtue of its effectiveness, need for less skilled man power and economical treatment of expansive soil.

With the addition of lime, the properties of the expansive soil change immediately due to the ion-exchange process and flocculation, making the soil more friable, less plastic and easy to work and compact (Diamond and Kinter 1965). The above reactions are termed as the short-term reactions or soil modification reactions. With the curing

J. S. McCartney and L. R. Hoyos (Eds.): GeoMEast 2018, SUCI, pp. 15–27, 2019.
https://doi.org/10.1007/978-3-030-01914-3_2

period, the lime treated samples attain improvement in the engineering properties through the soil-lime pozzolanic reactions. These soil-lime reactions are generally slower than the other calcium-based stabilizers, like the soil-cement hydration (CIRIA 1988, Osinubi and Nwaiwu 2006). Due to the pozzolanic reactions of the soil-lime mixture, new products of cementitious nature are formed, namely, the calcium silicate hydrate and the calcium aluminum hydrate, which improves the strength of the treated soil (Diamond and Kinter 1965).

Compacting the lime treated expansive soil immediately after mixing results in the reduction of the maximum dry density (MDD) and increase in the optimum moisture content (OMC) of the treated soil. The reduction in the MDD is small and averages at about 2.5% for most soils. The increase in the OMC is significant (by about 25% in some instances) with the addition of 2 to 3% lime (Mitchell and Hooper 1961). Addition of lime beyond 3% shows only a marginal increase in the OMC. Apart from the lime content, the compaction characteristics are also influenced by the type of lime used. Soils treated with quick lime have higher OMC than the soils treated with hydrated lime. However, the MDD is not affected by the type of lime used (Laguros and Davidson 1956, Minnick and Williams 1956). The compaction characteristics also depend on the reactivity of the lime, which in turn depends on the temperature and time of firing. Thus, the behaviour of lime treated soils depends on the origin and manufacturing process of the lime that is being used for the stabilization.

It is a known fact that the presence of sulfate in the expansive soil renders the lime stabilization ineffective in the service life of the treated soil. The presence of sulfate in the soil fabric results in the formation of expansive minerals, such as ettringite and thaumasite. These minerals render the treated soil to behave similar to that of the untreated expansive soils (Hunter 1988). It has been established that the soils containing the soluble sulfate more than 2000 ppm are termed as sulfate-rich soils or sulfate bearing soils (Mitchell and Dermatas 1990; Kota et al. 1996). The intrusion of sulfate into the soil occurs due to acid rains, underground water flow and moisture percolation due to evapotranspiration and construction water (Dermatas 1995). Although the presence of sulfate in the soil can be found in various forms, the most common among them are calcium sulfate, sodium sulfate and magnesium sulfate (Kota et al. 1996, Burkart et al. 1999, Puppala et al. 2003).

There are various theories that explain the formation of the ettringite and thaumasite in the soil fabric. But only two major theories of the formation of ettringite are widely accepted, namely, the crystal growth theory and water absorption or hydration theory (Mehta 1973a, Ogawa and Roy 1982, Mehta and Wang 1982, Min and Mingshu 1994). The formation of the ettringite takes a longer duration of time, more than 30 days, depending on the environmental conditions (Mitchell and Dermatas 1990; Diamond 1996). In addition to the long-term effects mentioned above, it has been proved that the presence of chlorites, carbonates, and sulfates acts as inhibitors to the soil modification reactions (Locat 1990). Due to this, the formation of the intermediate cementitious compounds during the lime stabilization may be retarded, which may result in a change in behaviour of lime treated soils, in particular, the compaction and strength characteristics. Hence, this study is aimed at understanding the effect of sulfate on the compaction characteristics of a lime treated expansive soil. The effect of time delay on the compaction characteristics and strength development of the sulfate contaminated lime treated expansive soil is also studied.

2 Materials

The sufficient quantity of expansive soil was collected at a depth of 2 m from a construction site at Tiruchirapalli, Tamil Nadu, India, for the present experimental research. The expansive soil was sieved through a 4.75 mm sieve to remove any gravel fraction, and then air-dried, pulverized and passed through a 425 μm sieve, and this soil was stored in an air-tight bin and used for the present research. The basic properties of the expansive soil are summarized in Table 1. The soil is classified as inorganic clay of high compressibility or fat clays (CH) based on the Unified Soil Classification System (USCS). The initial consumption of lime (ICL) value of the expansive soil, determined using Eades and Grim (1966) protocol, was found to be 3.5%.

Table 1. Properties of soil

Property	Value
pH	8.75
Conductivity (mS/cm)	2.77
Specific gravity	2.75
Grain size distribution (%)	
Sand	24.5
Silt	22.5
Clay	53
Atterberg limits (%)	
Liquid limit	95
Plastic limit	22
Shrinkage limit	7.5
Compaction characteristics	
Optimum moisture content (%)	20
Maximum dry density (Mg/m^3)	1.714
Differential free swell index (%)	300
IS Soil classification	CH
Initial consumption of lime (%)	3.5

The laboratory grade lime ($CaOH_2$) and sodium sulfate (Na_2SO_4) with a purity of 90% and 99%, respectively, were used for the present study. Sodium sulfate was chosen as it has the maximum solubility of 49.7 mg/100 mL among all the common sulfate salts available.

3 Experimental Procedures

The expansive soil passing through 425 μm is mixed with various percentages of lime (2.5, 3.5, and 4.5%) by dry weight of soil in the dry state. The percentages of lime were fixed based on the lime fixation point i.e. one value corresponds to the ICL value (3.5%), one value corresponds to below ICL (2.5%) and the other one above the ICL (4.5%). The required amount of water was added to the dry soil-lime mix and mixed with the help of spatulas to carry out the mini Proctor compaction tests (Sridharan and Sivapullaiah 2005). Immediately after mixing the required quantity of water, the compaction tests were carried out without any time delay at various water contents and the compaction curves were obtained. In order to determine the effect of compaction delay on the various properties of lime treated soil, the soil was mixed with the required percentage of lime in the dry state. The predetermined amount of water, corresponding to the optimum moisture content, was added to the soil-lime mix, and then this wet mix was left undisturbed for periods of 0 h (no delay), 6 h and 24 h. The selected time delay replicates the field situation. During the mellowing period, enough measures were taken to avoid the loss of water from the soil-lime mixes by placing them in plastic zip-lock bags and equilibrated in desiccators.

In order to find the effect of sulfate (as a pore fluid) on the compaction and strength behaviour of the lime treated expansive soil, the sodium sulfate solutions of 5000, 10000 and 20000 ppm, designated as S5, S10, and S20, respectively, were mixed separately with the soil-lime mixtures. The concentrations of sulfate solutions used in the present study represent the general concentration of effluents from the tannery industries and the concentration of sulfate in acid rains. Immediately after the mixing with the required quantity of sulfate solution of desired concentration, the compaction tests were carried out. The same protocol was adopted for finding the effect of compaction delay on the sulfate contaminated soil-lime mixtures.

The unconfined compressive strength was determined on specimens of size 38 mm diameter and 76 mm height. These specimens were prepared by static compaction to their respective maximum dry density at corresponding optimum moisture content. In order to bring out the effect of compaction delay, the specimens were compacted after allowing the desired time delay (mellowing period). The UCC specimens thus prepared were cured at a relative humidity of 100% for various curing periods before testing. The effect of curing on the treated specimens was brought out by adding DW or sulfate solution to saturate in a sand bath. The following steps were employed for the preparation of sand bath (Fig. 1). First, 10 mm layer of sand was placed at the bottom of the bath and leveled. After which the specimens were placed in the tub so that none of the specimens interfere with each other. This was followed by addition of sand filling the gaps between the specimens and extra sand was added up to a level of 20 mm above the height of the specimens and leveled. The loading plate was placed above the sand layer and above which a nominal load of 6.25 kPa was placed. The specimens were taken out after the desired period of curing for carrying out the UCC tests.

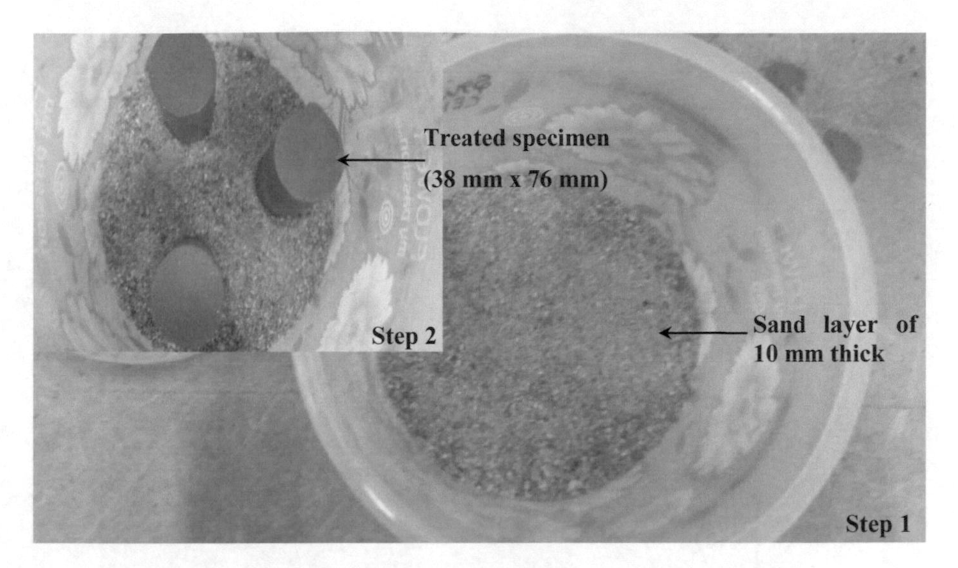

Fig. 1. Sand bath arrangement for curing of lime treated specimens

4 Results and Discussion

Figures 2a, b and c bring out the effect of sulfate concentration on the compaction behavior of the lime treated expansive soil with lime contents of 2.5, 3.5 and 4.5%, respectively. It can be observed from Fig. 2 that the addition of lime to the expansive soil resulted in an increase in the OMC from 24.4% to 25.5%. On further increase in the lime content from 2.5% to 3.5%, the OMC increased from 25.5% to 26.5%. The MDD of the lime treated soil decreased to 1.583 Mg/m^3 from 1.611 Mg/m^3 for an increase in the lime content from 2.5% to 4.5%. From the above results, it can be observed that the addition of lime to the expansive soil results in an increase in the OMC along with a decrease in the maximum dry density. Similar results were reported by Bell (1996) and Sridharan and Prakash (1998).

The addition of sodium sulfate to the lime treated soil resulted in a further decrease in the MDD. However, the OMC increased with respect to the untreated soil OMC, but decreased with respect to the lime treated soil OMC. The addition of 5000 ppm sulfate solution to 2.5% lime treated soil resulted in a decrease in the MDD to 1.59 Mg/m^3 from 1.611 Mg/m^3. The same trend of decrease in the MDD was observed across the various lime contents. From Fig. 2, it is evident that the compaction curves of the sulfate contaminated soils are flatter when compared to the untreated and lime treated soils. However, the variation of the sulfate concentration from 5000 to 20000 ppm had a negligible effect on the MDD and OMC of the lime treated expansive soil.

Figures 3a and b compare the compaction curves of the uncontaminated lime treated expansive soil with the sulfate contaminated lime treated soil corresponding to compaction delay of 6 h and 24 h, respectively. From the figures, it can be observed that as the lime content increased, the MDD of the lime treated soil decreased. In case of sulfate contaminated lime treated soils, the MDD increased with respect to the

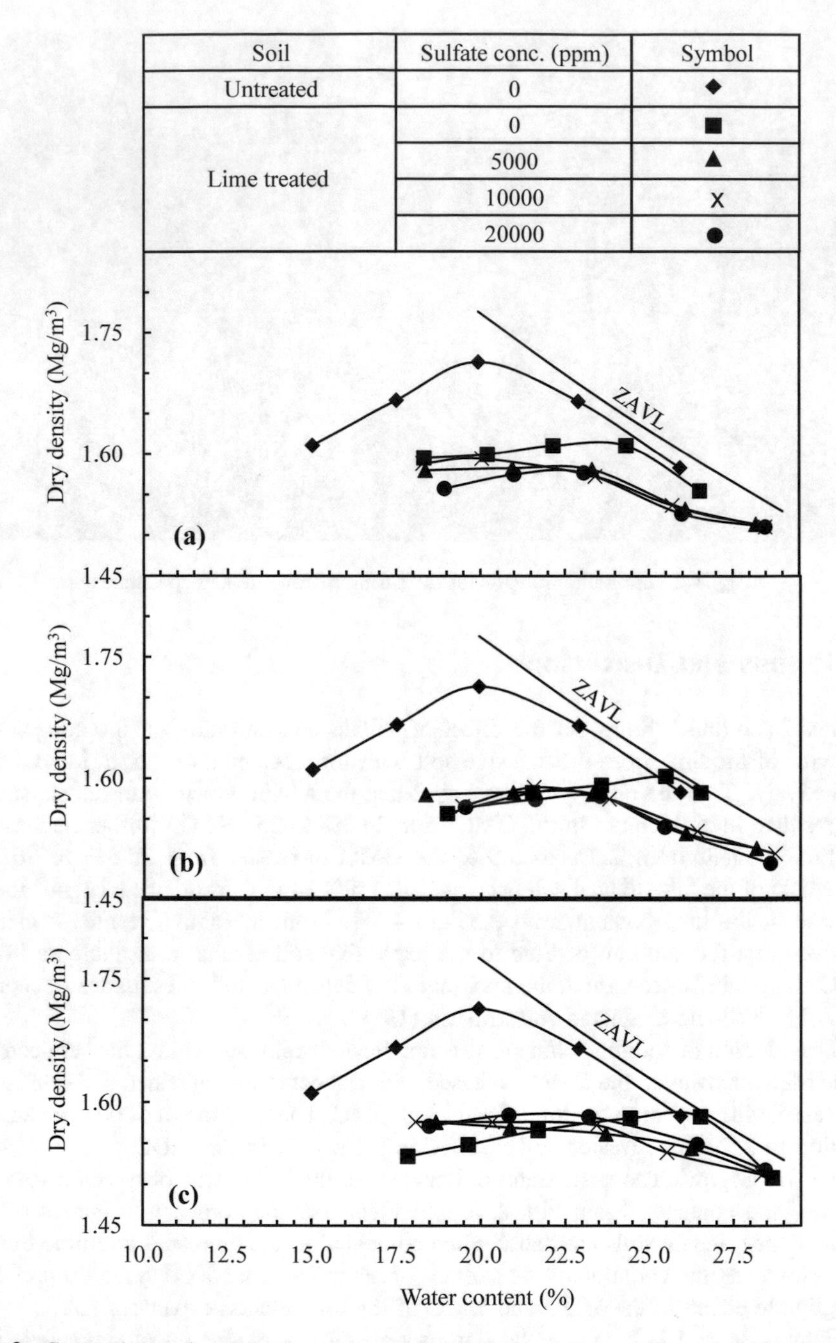

Fig. 2. Comparison of compaction curves of soil-lime mixtures with and without sulfate contamination at lime content of: (a) 2.5% (b) 3.5% and (c) 4.5%

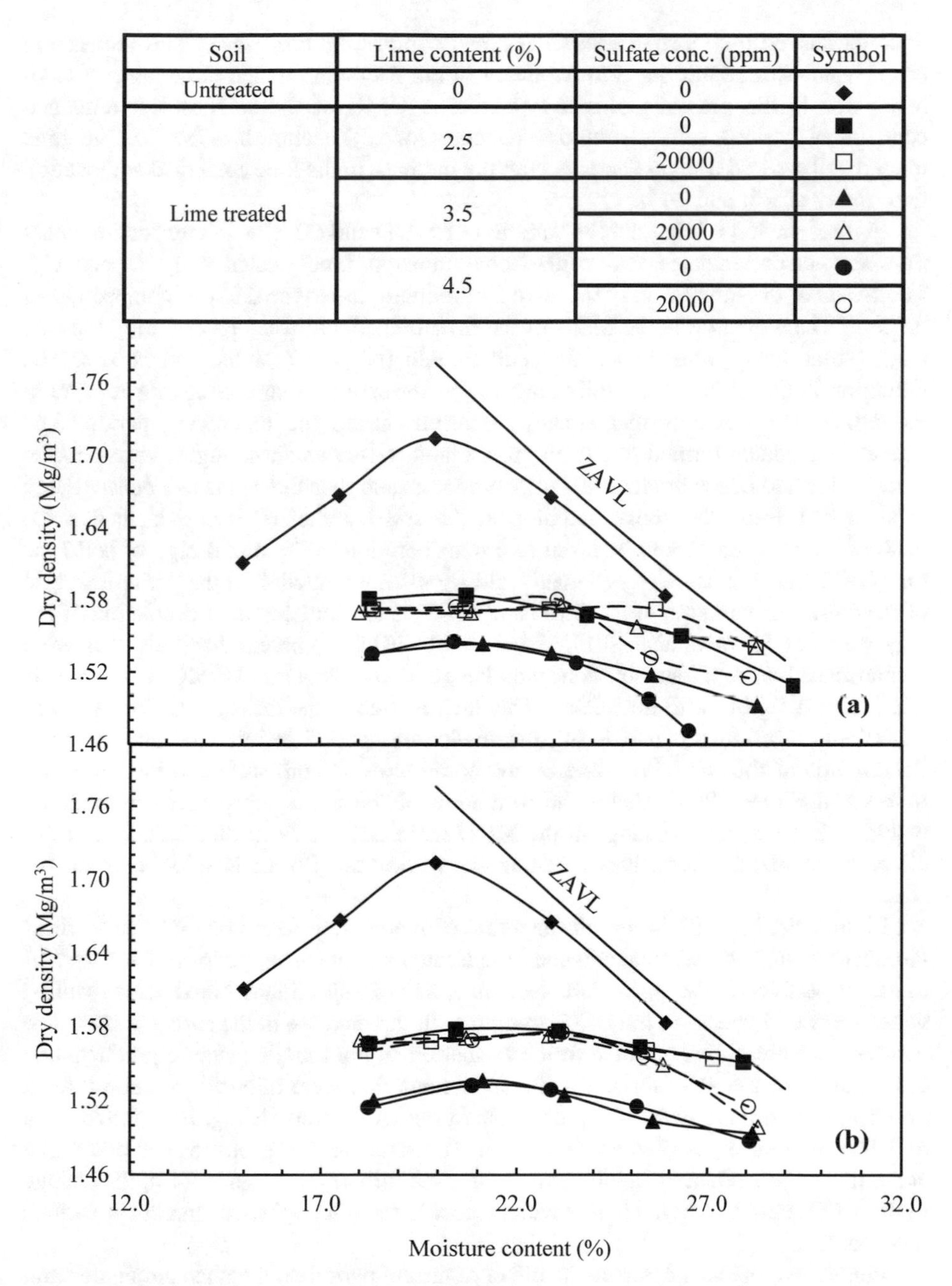

Fig. 3. Comparison of compaction curves of untreated and lime treated soil (with and without sulfate contaminated) with a compaction time delay of: (a) 6 h (b) 24 h

uncontaminated lime treated soils at identical compaction time delay. This increase in MDD value can be attributed to the delay in the formation of the pozzolanic components due to the presence of sulfate salt. The OMC of the soils almost remained constant at various sulfate solution concentrations. Whereas the OMC of the lime treated soil showed a slight decrease with the increase in the lime content at compaction time delay of 6 h and 24 h.

Figures 4a and b bring out the variation of MDD and OMC with compaction delay time for uncontaminated and sulfate contaminated lime treated soil, respectively. The MDD of the lime treated expansive soil reduces due to the delay in compaction of the soil. The reduction in the MDD of the lime treated soil was higher during the first 6 h of time delay, after which the reduction in the MDD of the soil is less. The reduction in the MDD of the soil is due to the formation of coarser aggregates due to the pozzolanic reactions that already occurred during the mellowing period. The coarser aggregates formed due to the pozzolanic reactions show higher resistance to compaction and hence produce larger macropores, and thus the soil is less dense (Sante et al. 2015). From the above variation in the soil parameters, it is evident that the delayed compaction should be taken into consideration in the mix design of soil-lime mixtures because it causes a systematic and significant reduction of the dry unit weight of the soil. This variation in the OMC and MDD is attributed to the flocculation of the clay particles (Sivapullaiah 2011, Sante et al. 2015). Whereas for soils that were contaminated with sulfate solution, the change in the MDD and OMC of the soil is relatively negligible with time delay. This may be due to the fact that the formation of pozzolanic reaction product during the mellowing period is inhibited and the only flocculation of the soil takes place by the addition of sodium sulfate to the soil (Diamond and Kinter 1965). Hence the formation of the aggregate particles is retarded resulting in a negligible change in the MDD and OMC of the treated soil. The above effect is visualized irrespective of the quantity of lime stabilizer added to the expansive soil.

Figures 5a, b, and c bring out the effect of mellowing period on UCC strength of the uncontaminated and contaminated lime treated soil at curing periods of 1, 7 and 14 days, respectively. The uncontaminated lime treated specimens cured with distilled water show an increase in the UCC strength with the increase in the curing period. The increase in strength is attributed to the formation of pozzolanic reaction products and cementation of the soil fabric. For the specimens that were allowed to mellow for a duration of 6 h, there was a sharp decrease in the UCC strength, e.g. from 0.973 MPa to 0.746 MPa for the 2.5% lime treated soil. The same trend was observed at other lime contents as well. When the mellowing period was further increased to 24 h, the change in the UCC strength was negligible with respect to the 6 h mellowed strength at various lime contents.

Figures 6a, b, and c compare the effect of curing period on the uncontaminated and sulfate contaminated lime treated specimens prepared after mellowing period of 1, 6 and 24 h, respectively. From Fig. 6a, it is evident that the UCC strength of uncontaminated lime treated soil is higher than the sulfate contaminated lime treated soil at all the curing periods for a mellowing period of 1 h. For example, the strength of 3.5% lime treated uncontaminated soil mellowed for 1 h is 1.189 MPa, 2.14 MPa, and 2.51 MPa for specimens cured for 1, 7, and 14 days, while the strength of the

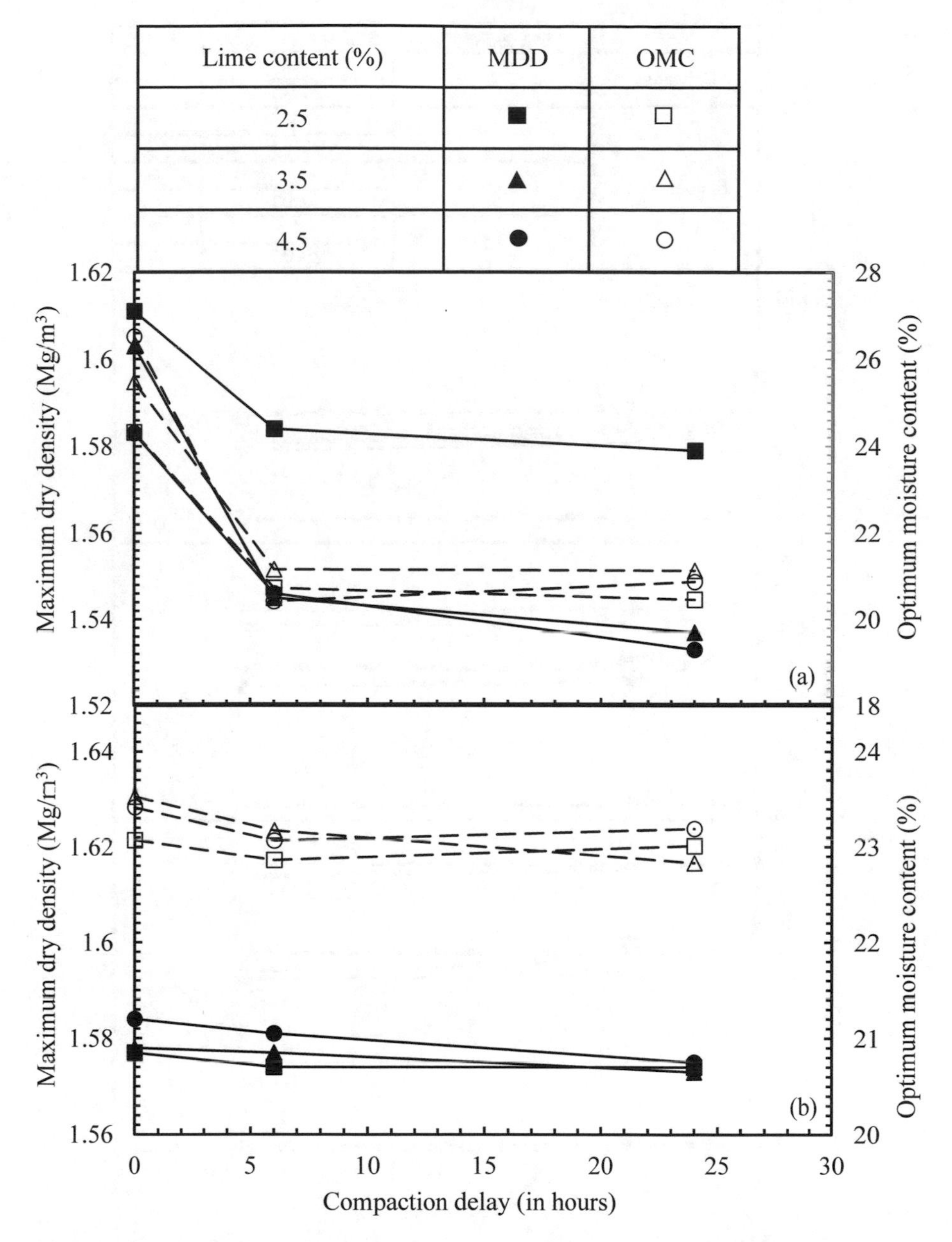

Fig. 4. Variation of maximum dry density and optimum lime content with compaction time delay for: (a) lime treated expansive soil (b) sulfate (20000 ppm) contaminated lime treated soil

specimens after placing under S20 solution for the same curing periods resulted in lower values of strength – 0.964 MPa, 1.96 MPa, and 2.29 MPa, respectively. While for the specimens that were mellowed for 6 h and 24 h, the UCC value of the sulfate contaminated lime treated specimens were higher than the uncontaminated lime treated

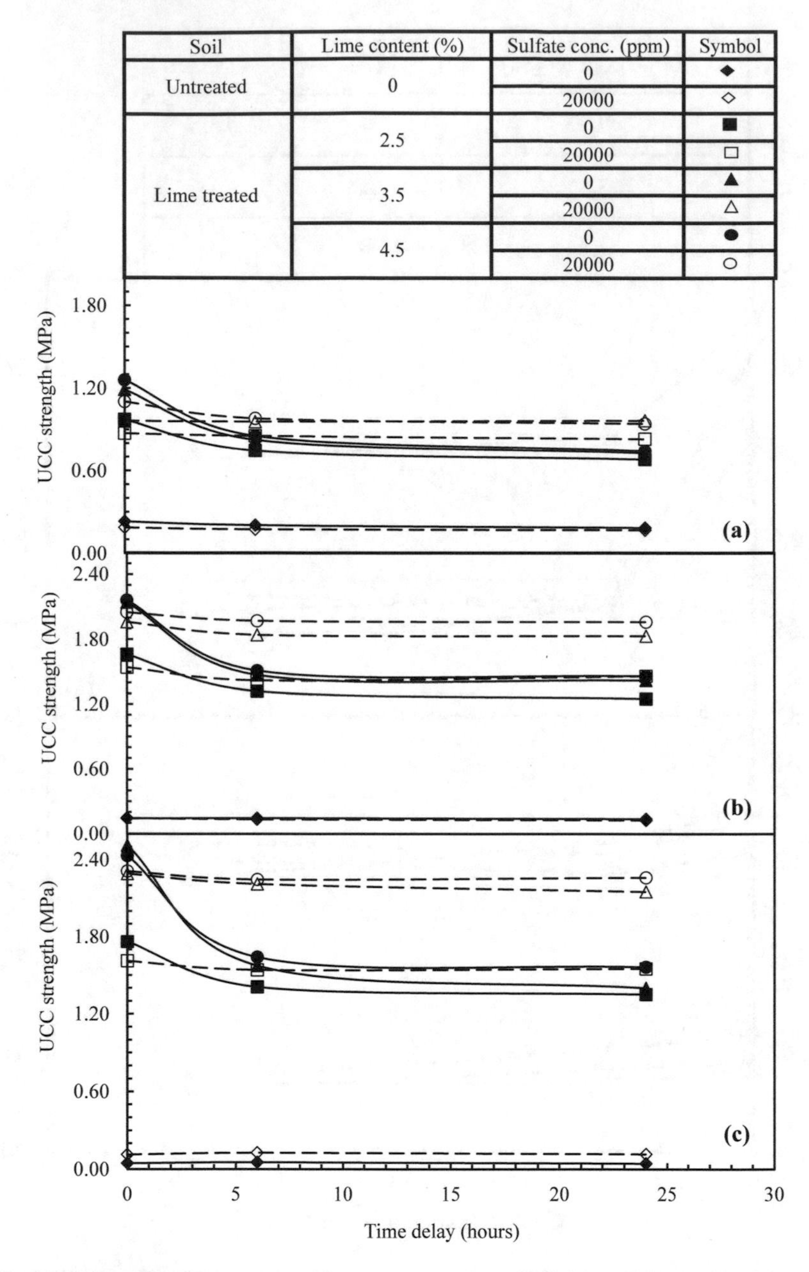

Fig. 5. Variation of UCC strength with compaction time delay for soil-lime mixtures with and without sulfate contamination cured for: (a) 1 day (b) 7 days and (c) 14 days

specimens. For example, the strength of 3.5% lime treated uncontaminated soil mellowed for 6 h is 0.824 MPa, 1.47 MPa and 1.573 MPa for specimens cured for 1, 7, 1st 14 days, while the strength of the specimens contaminated with S20 solution for the

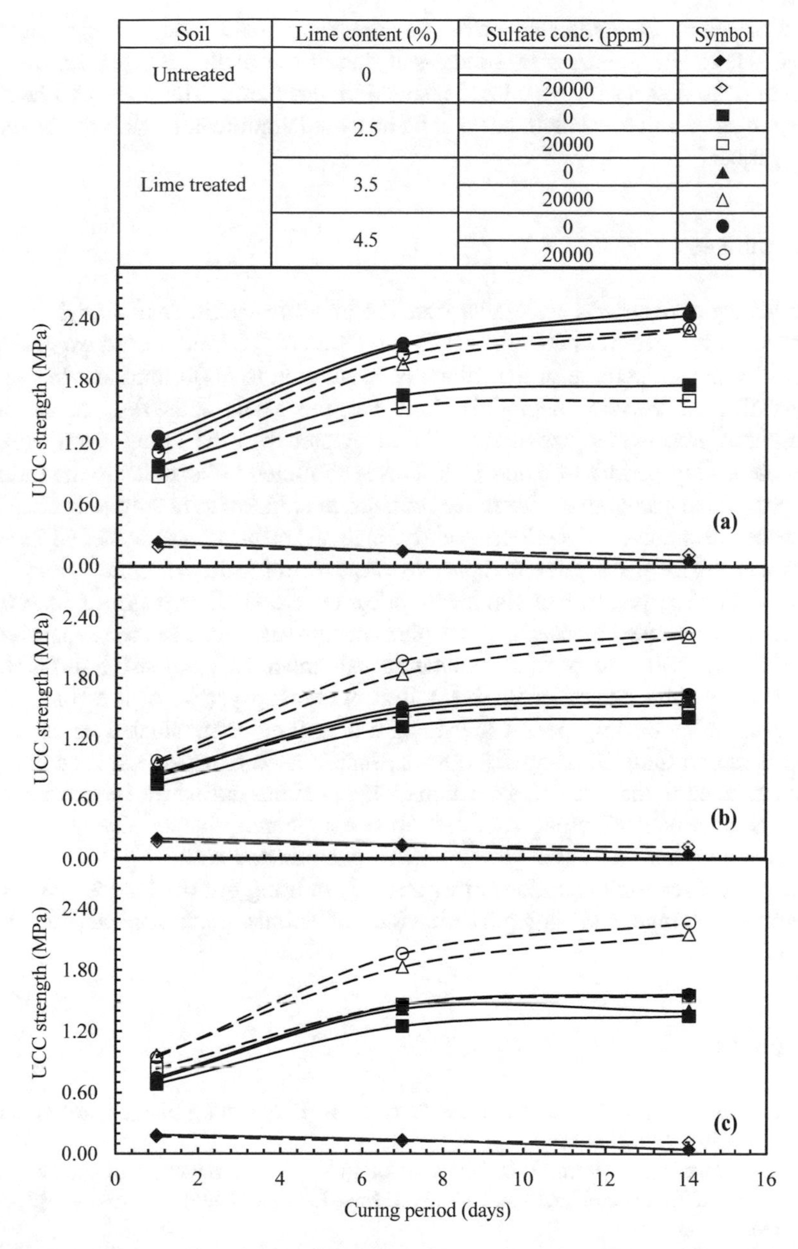

Fig. 6. Variation of UCC strength with curing period for soil-lime mixtures with and without sulfate contamination at compaction time delay of: (a) 0 h (b) 6 h and (c) 24 h

same curing periods resulted in higher UCC values of 0.957 MPa, 1.84 MPa and 2.21 MPa, respectively. The above behavior of the expansive soil is attributed to the formation of aggregations in the lime treated soil during the mellowing period, due to which the macropores size increased and resulted in reduced strength of the soil.

In case of sulfate contaminated soils, the formation of the flocs or aggregations is retarded due to the presence of sulfate salt during the mellowing period, and upon compaction the density increased and resulted in less porous structure, and hence the UCC strength of sulfate contaminated soils increased significantly with the increase in curing period.

5 Conclusions

The following conclusions are drawn from the present experimental study:

At mellowing period of 0 h, the MDD and OMC of the lime treated expansive soil decreased with the increase in the sulfate concentration to 5000 ppm and beyond this concentration the effect is negligible. However, the MDD and OMC of the sulfate contaminated lime treated expansive soils are greater than the lime treated expansive soils at mellowing periods of 6 and 24 h. This is attributed to the delay in the formation of the pozzolanic compounds due to the presence of sulfate salt in sulfate contaminated lime treated expansive soils. Consequently, higher densities were achieved in sulfate contaminated soils at the same compactive effort owing to lesser aggregations.

The mellowing period had significant effect on the UCC strength of lime treated soils, whereas the effect is marginal in sulfate contaminated lime treated expansive soil. Further, the UCC strength of sulfate contaminated lime treated expansive soil is slightly lower than the lime treated expansive soil at mellowing period of 0 h, but with the increase in the mellowing period the strength of sulfate contaminated expansive soil increased substantially in comparison to the lime treated expansive soil counterparts. This is attributed to the retarded formation of aggregations during the mellowing period due to the presence of sulfate salt, and upon compaction the density increases and results in the higher UCC strength of sulfate contaminated soils.

The present research calls for further research to bring out the long term effects on both volume change and strength behaviour of sulfate contaminated lime treated expansive soil.

References

Bell, F.G.: Lime stabilization of clay minerals and soil. Eng. Geol. (1996). https://doi.org/10.1016/0013-7952(96)00028-2

Burkart, B., Goss, C.G., Kern, P.J.: The role of gypsum in production of sulfate induced deformation of lime-stabilized soils. Environ. Eng. Geosci. (1999). https://doi.org/10.2113/gseegeosci.v.2.173

Diamond, S.: Delayed ettringite formation-processes and problems. Cement Concrete Compos. PII:SO958-9465(96)00017-O

Diamond, S., Kinter, E.B.: Mechanism of soil-lime stabilization. U.S. bureau of public roads. Highw. Res. Board **92**, 83–102 (1965)

Hunter, D.: Lime-induced heave in sulfate bearing clay soils. J. Geotech. Eng. ASCE (1988). https://doi.org/10.1061/(asce)0733-9410(1988)114:2(150)

Kota, P.B., Hazlett, D., Perrin, L.: Sulfate-bearing soils: problems with calcium based stabilizers. Transp. Res. Rec. J. Transp. Res. Board (1996). https://doi.org/10.3141/1546-07

Laguros, J.G., Davidson, D.T.: Evaluation of lime for stabilization of loess. Proc. ASTM **56**, 1301 (1956)

Mehta, P.K.: Mechanism of expansion associated with ettringite formation. Cem. Concr. Res. (1973a). https://doi.org/10.1016/0008-8846(73)90056-2

Min, D., Mingshu, T.: Formation and expansion of ettringite crystals. Cem. Concr. Res. (1994). https://doi.org/10.1016/0008-8846(94)90092-2

Minnick, L.J., Williams, R.: Field evaluation of lime-fly ash-soil composition for roads. Highw. Res. Board Bull. **129** (1956)

Mitchell, J.K., Hooper, D.R.: Influence of time between mixing and compaction on properties of a lime stabilised expansive clay. Highw. Res. Board Bull. 304 (1961)

Ogawa, K., Roy, D.M.: C_4A_3S hydration ettringite formation and its expansion mechanism: iii. effect of CaO, NaOH and NaCL; conclusions. Cem. Concr. Res. (1982). https://doi.org/10.1016/0008-8846(82)900-114

Osinubi, K.J., Nwaiwu, C.M.O.: Compaction delay effects on properties of lime-treated soil. J. Mater. Civ. Eng. ASCE (2006). https://doi.org/10.1061/(asce)0899-1561(2006)18:2(250)

Puppala, A.J., Wattanasanticharoen, E., Hoyos, L.: Ranking of four chemical and mechanical stabilization methods to treat low-volume road sub grades in Texas. Transp. Res. Board (2003). https://doi.org/10.3141/1819b-09

Ramesh, H.N., Sivapullaiah, P.V.: Role of moulding water content in lime stabilisation of soil. Ground Improv. (2011). https://doi.org/10.1680/grim.900040

Sante, M.D., Fratalocchi, E., Mazzieri, F., Brianzoni, V.: Influence of delayed compaction on the compressibility and hydraulic conductivity of soil-lime mixtures. Eng. Geol. (2015). https://doi.org/10.1680/grim900040

Sridharan, A., Prakash, K.: Mechanism controlling the shrinkage limit of soils. Geotech. Test. J. (1998). https://doi.org/10.1520/gtj10897j

Sridharan, A., Sivapullaiah, P.V.: Mini compaction test apparatus for fine grained soil. Geotech. Test. J. (2005). https://doi.org/10.1520/gtj12542

Combined Effects of Bottom Ash and Lime on Behaviour of Expansive Soil

Thang M. Le[1(✉)], Liet C. Dang[1], and Hadi Khabbaz[2]

[1] School of Civil and Environmental Engineering, University of Technology Sydney (UTS), Sydney, Australia
minhthang.le@student.uts.edu.au

[2] Geotechnical Engineering, School of Civil and Environmental Engineering, University of Technology Sydney (UTS), Sydney, Australia

Abstract. This study illustrates the effectiveness of combining bottom ash and hydrated lime to enhance the engineering properties of expansive soil. The bottom ash was collected from Eraring Power Station in New South Wales, Australia, as a by-product of coal-fired power stations, and soil specimens were used as artificial soil including kaolinite, bentonite and fine sand in a reasonable ratio to stimulate soil samples with characteristics of expansive soil. The stabilised soil samples were prepared by altering the bottom ash content from 0% to 30% on a dry weight basis of expansive soil as well as with constant percentage of 5% in hydrated lime. Through conducting a series of experimental tests including linear shrinkage and unconfined compressive strength (UCS) in various curing time, the shrinkage and strength behaviour of treated soils were investigated and compared with untreated soil samples. The results revealed that the combination of bottom ash and hydrated lime significantly reduced the linear shrinkage, while it increased the strength of expansive soil. The use of bottom ash alone is not recommended due to a slight increase of linear shrinkage and a minor negative impact on the soil strength. The optimum content of combined bottom ash and hydrated lime to stabilise expansive soils is also presented.

Keywords: Bottom ash · Expansive soil · Hydrated lime · Linear shrinkage Unconfined compressive strength

1 Introduction

Expansive soil is considered as a pivotal factor that poses threat to foundation of many civil constructions, such as highway or embankments. The soil has a high potential of swelling and shrinkage under the changes of ambient moisture. The displacement range of changing from its swell to collapse can be considerable and expand the gaps in the unsaturated soil ground, causing the additional pressure developed in civil structures above. The phenomenon with increasing cracks becomes obvious to observe once the stress increases to the stage where the strength capacity of highway subgrade surface or slab foundation is over, causing heave or settlement. A number of damage induced by expansive soil is reported in many countries, where the expansive soil occupies (Phani Kumar and Sharma 2004). The expense for building maintenance caused by ground

J. S. McCartney and L. R. Hoyos (Eds.): GeoMEast 2018, SUCI, pp. 28–44, 2019.
https://doi.org/10.1007/978-3-030-01914-3_3

expansion or collapse is always costly and requires an astronomical cost to repair or even reconstruct. For example, the annual statistics from damage cases in USA and UK indicates that there was a cost of over 15 billion dollars and about 400 million pounds by the impact of expansive soil on foundations of the infrastructure (Gourley et al. 1993; Jones and Jefferson 2012; Nelson and Miller 1997; Viswanadham et al. 2009). Therefore, it is necessary to control the swelling-shrinkage behaviour of expansive soil to limit its detrimental impact.

However, the degree of swelling or shrinkage of expansive soil is not consistent and varies from one region to another. Even in the same location, presence of expansive soil makes the soil characterisation harder. This type soil is also known as problematic soil. Nevertheless, for studying the behaviour of this soil systematically, it can be noted that some chemical compounds in this soil play a primary role for its swelling and shrinking property. Of the chemicals considered, montmorillonite, a kind of expansive clay mineral, can speed up the swelling and shrinkage rate, whereas kaolinite is perceived as a less expansive material, which helps to reduce the swelling incidence (Das and Sobhan 2013; Holtz et al. 2011). Likewise, sand as a coarse and non-plastic material can diminish the swelling potential of expansive soil. The combination of three materials can produce a compacted sample that illustrates the field characteristic of expansive soil, and thus a desired constitutive sample can be formed, which facilitates research on the complex behaviour of soft soil (Le 2015; Le et al. 2015).

The existing methods to limit the problematic behaviour of expansive soil, are stone columns (Fatahi et al. 2012), pile supported and geosynthetic earth platform (Han and Gabr 2002; Liu et al. 2007), sand cushion technique (Satyanarayana 1996), belled piers (Chen 1988), granular pile-anchors (Phani Kumar 1997) and chemical stabilization (Erdal 2001, Obuzor et al. 2012). Among the above-mentioned methods, the chemical treatment, mixing chemical agents with soil, is more promising and received a great deal of attention through extensive studies (Bergado et al. 1996; Erdal 2001; Edil et al. 2006; Fatahi et al. 2012, 2013; Khabbaz and Fatahi 2012; Nguyen et al. 2014; Osinubi et al. 2009; Phani Kumar 2009). The binders for soil chemical stabilization can be originated from variety of sources, but simply they can be classified into two main categories, including agricultural and industrial materials. For agricultural by-products, some ashes like bagasse ash were explored to enhance the engineering properties of expansive soil (Dang et al. 2015, 2016; Osinubi et al. 2009). Other farming wastes, such as rice husk ash, coconut coir fibre or even eggshells, are utilized to enhance the soil strength and alleviate the problems related to free swelling and shrinkage of expansive soil (Vivi et al. 2015; Basha et al. 2003; Rahman 1986; Sivakumar Babu et al. 2008). Meanwhile, industrially originated products, such as fly ash, silica fume or cement, become common stabilizing additives in soil treatment (Iman and Schoobbasti 2003; Consoli et al. 1998; Lorenzo and Bergado 2004; Miller and Azad 2000, Rifai et al. 2010).

Hydrated lime is also an outstanding example for treatment of expansive soil because of its impressive improvement effect on soil strength. The development of strength in lime-soil mixture can be clarified into two processes, namely, the modification and stabilization (Nguyen et al. 2014). While soil modification relates to reactions of flocculation and cation exchange between lime and clay in short term, the next process in long terms is involved in the soil stabilisation with pozzolanic reactivity to form cementitious bonds with crystal formulation, namely compounds-calcium silicate hydrates (CSH) and calcium aluminate hydrates (CAH) in the soil blend. Further studies on the combination of lime and bagasse ash in soil admixture have proven the better enhancement in soil strength by addition of divalent and trivalent cations, such as Ca^{2+}, Fe^{3+} and Al^{3+}, from ash to enhance exchange of cation with reactive soil (Chen 1988; Ganesan et al. 2007; Goyal et al. 2007; Manikandan and Moganraj 2014; Osinubi et al. 2009; Sharma et al. 2008). Bottom ash is one of such pozzolanic materials that can be utilized to improve this exchange formulation. However, there are very limited studies on the combination of bottom ash and lime for stabilisation of expansive soil. Hence, further investigation on their combined dosage for soil treatment is necessary to provide better understanding of the bottom ash-lime-soil behaviour.

Bottom ash is a by-product from the burning process in coal-fired power stations, which constitutes most production of coal ash to produce energy. Unlike the fly ash with fine structure, bottom ash is a granular and coarse material. Therefore, it is reasonable for treatment of fine-grained soil, in applications related to geotechnical solutions for fill materials with large volumes, such as highways, embankments, fills or backfills (Kim and Prezzi 2008). Therefore, bottom ash can produce high quality aggregates when it is combined with fine clay binders (Geetha and Ramamurthy 2011). Furthermore, the coal ash has an extensive record in utilization for soft soil stabilization due to its pozzolanic reactions and self-cementing properties (Kayabali and Bulus 2000; Mackiewicz and Ferguson 2005). The engineering performance of bottom ash as a construction material, relies on various factors, such as the density, the grain size distribution, the compaction properties, the hydraulic conductivity and the shear strength parameters (Cheriaf et al. 1999; Huang and Lovell 1990; Jorat et al. 2011; Kim and Do 2012; Kim et al. 2005; Kim et al. 2011). From these influencing factors, recent studies on bottom ash have indicated negative effects on soil compression strength when only bottom ash is used to stabilize expansive soil without combination of other activators like basanite or rice husk ash (Kamei et al. 2013, Modarres and Nosoudy 2015). In contrast, a noticeable increase in strength is observed in soil mixtures with lime/cement and bottom ash because of the cementing ability of hydrated lime/cement and pozzolanic impact of bottom ash (Geliga and Ismail 2010; Kolay et al. 2011; Kayabali and Bulus 2000; Mackiewicz and Ferguson 2005).

Annually about 2.5 billion tons of coal ash in the world is originated from mines and destined to landfills as waste material (Doulati Ardejani et al. 2010; Garcia et al. 2012). It might include 1.5 million tons of coal ash waste in Japan reported in 2000 (Kunitomo 2009), or 6.9 million tons in Australia which stockpiled in compacted fills (Lav and Kenny 1996). The combination of bottom ash and conventional stabilisers such as lime and cement to treat soil increased the admixture strength and improved its durability (Kinuthia and Nidzam 2009; Lu et al 2014; Obozor et al. 2012; Shibi and Kamei 2014). Therefore, the bottom ash-lime combination for a solidification soil mixture becomes not only a great interest of many researchers for chemical stabilization of expansive soil but also a promising solution to minimise the issues of disposing bottom ash in terms of environmental protection (Kim and Prezzi 2008).

In Australia, although there were comprehensive studies on fly ash as a siliceous additive for soil embankment or pavement-subgrade materials, few papers have been published on the utilization of bottom ash as a soil stabilizer for the stabilization of expansive soil. In this paper, an array of tests have been performed on untreated and treated expansive soil samples in conjunction with various contents of bottom ash and hydrated lime after curing durations of 7 and 28 days. The experimental tests, including linear shrinkage and unconfined compressive strength (UCS) tests, are employed to investigate the combined effect of hydrated lime and bottom ash on the shrinkage potential and the compressive strength of expansive soil.

2 Materials

2.1 Expansive Soil

In this paper, a mixture of Kaolinite Q38, Bentonite Active Bond 23 and Sydney sand in fine size (KBS) was employed to obtain the reconstituted samples. The bentonite with darker grey colour has the higher linear shrinkage (LS) and liquid limit (LL) than Kaolinite (35% as against to 9% in LS, and 340% compared to 50.5% in LL, respectively). While the LS testing procedure is followed the Australia Standard (AS) 1289.3.4.1 (2008), the Atterberg limit values were determined based on AS 1289.3.9.1 (2015). To simulate the expansive soil found in Queensland, Australia, uniformly graded sand was mixed with kaolinite and bentonite at a fixed proportion of 65% kaolinite, 30% bentonite and 5% fine sand by the total dry mass. The combination ratio of admixture was obtained after various trials of different portions to achieve a LS value of 21.7% identical to that of Queensland expansive soil, according to the recent study on this problematic soil (Dang et al. 2015, 2016, and 2017). Once the LS value of 21.7% was achieved, the liquid limit of the blend was measured at 155%. Meanwhile, the plastic limit was obtained at 30.92%, resulting in a high value of Plastic Index (PI) at 129%. With these high LS and PI values, the reconstituted soil can be classified as highly expansive soil. In term of particle size, the reconstituted soil can be regarded as high compressive clay (CH), according to ASTM D2487 (2011). The detailed index properties of the soil are shown in Table 1.

2.2 Bottom Ash

Bottom coal ash utilized in this study were obtained from Eraring Power Station, New South Wales, Australia. The ash is classified as class F fly ash with grey colour and low calcium content. The bottom ash has the natural moisture content at 25.08% in average and specific gravity (G_s) of 1.96. Since its plastic limit is unable to be determined, this ash can be classified as a non-plastic material. Meanwhile, from the result of sieving analysis, the particle size distribution curve of bottom ash indicates that the ash can be considered as poorly graded sand (SP) with C_u = 24 and C_c = 0.38, in accordance with ASTM D2487 (2011). Furthermore, with zero LS value, the ash material is expected to reduce the shrinkage potential of testing soil once it is added to soil samples. More details about characteristics of bottom ash are provided in Table 2.

Table 1. Mechanical properties of constitutive expansive soil

Characteristics	Value
Gravel	0.00
Sand	7.85
Silt/Clay	74.62
Liquid limit (%)	154.95
Plastic limit (%)	30.92
Plasticity index (%)	124.03
Linear shrinkage (%)	21.23
Specific gravity	2.64
Optimum moisture content (%)	29.04
Maximum dry density (kN/m^3)	13.23
USCS Classification of the soil	CH

Table 2. Physical and chemical characteristics of bottom ash

Physical properties		Chemical properties*	
Properties	Value	Components	% by weight
Gravel (%)	39.60	SiO_2	64.65
Sand (%)	54.15	Al_2O_3	24.7
Silt/clay (%)	6.25	Fe_2O_3	3.49
Specific gravity	1.96	CaO	1.61
Liquid limit (%)	67.94	MgO	0.78
Plastic limit (%)	–	SO_3	0.05
Optimum moisture content (%)	17.00	Na_2O	0.58
Maximum dry density (kN/m^3)	9.48	K_2O	1.59
Linear shrinkage (%)	0.00	TiO_2	0.96
USCS Classification	SP	LOI	1.28

Note: * after Lav and Kenny (1996)

2.3 Hydrated Lime

Hydrated lime used in this paper is a commercial product, which was manufactured in Adelaide in Australia and supplied by Cement Australia. The lime contains 75–80% calcium hydroxide and 7% silica in term of dry weight. The lime powder was kept in tight bags to purposely avoid its contact with ambient humidity.

3 Testing Program

3.1 Mixing Procedure

In this investigation, the additive contents in the admixture are attained by the ratio of their dry weight to the dry weight of soil. Artificial expansive soil with particles size smaller than 2.36 mm was prepared to be mixed with bottom ash or hydrated lime in the percentage by the dry weight of soil, which is shown in Table 3.

Table 3. Summary of testing mixtures

Mix No.	Bottom ash (%)	Hydrated lime (%)
1	0	0
2	5	0
3	10	0
4	20	0
5	30	0
6	0	5
7	5	5
8	10	5
9	20	5
10	30	5

3.2 Linear Shrinkage Test

In this testing method, an amount of soil specimen of about 250 g, passed the 425 μm sieve in dry state, were used for the test, following the procedure of sample reparation in AS 1289 for liquid limit and linear shrinkage tests. The linear shrinkage tests were performed on both untreated and bottom ash treated soil samples without or with hydrated lime. The linear shrinkage testing method specified in Australia standard (AS) 1289.3.4.1 (2008) was abided to determine the linear shrinkage values of treated soils.

3.3 Unconfined Compression Test

Adhering to AS 5101.4 (2008) standard, the unconfined compression (UC) test was performed in the specimens of treated and untreated expansive soils. For treated samples, the reconstituted soil was mixed with bottom ash and/or hydrated lime at designed contents before adding water to attain the optimum moisture content (OMC). The OMC value was obtained from the compaction test of untreated expansive soil, and its value was then applied to all UC samples. By applying the same OMC value of untreated soil, the maximum dry densities (MDD) of soils treated with different additive contents were determined by a series of many compaction tests prior to tamping additive-soil mixtures into a steel mould of 50 mm in diameter and 100 mm in height. The inner wall of mould was coated with a thin layer of grease to minimize the disturbance effect on samples, which can occur in the process of sampling extrusion. To obtain the desired dry density using the tamping technique in the cylindrical mould, the specimen was compacted in three equal layers to ensure that the specimen was uniformly compacted. After that, the samples were extruded, sealed in plastic wrap and cured for various durations of 7, 28 and 56 days in a temperature-controlled room at 25 °C and relative humidity at 80%. The periods of curing were noted to unwrap the samples at specific testing dates for unconfined compression tests. Before testing, the samples were weighed and their dimensions were also measured to recalculate and check the value of dry density. The loading frame Tritech with the capacity of 50 kN was employed in the UC test, including a linear variable differential transformer (LVDT) and S-shaped load cell. These two devices were indirectly connected to a computer via a data logger, which displays the dial data to users by an interface of Datacomm program. Saved in separated Excel files, the data was presented in columns of vertical displacement from LVDT and force from load cell. The strain rate was approximate 1 mm/min, and unconfined compressive strength (UCS) was obtained by the division of force to the equivalent circular area of the samples. The same test for every admixture was repeated three times to obtain their final average strength.

4 Results and Discussion

4.1 Linear Shrinkage

Figure 1 shows the effect of bottom ash on linear shrinkage after curing duration of 7 days. As can be seen in Fig. 1, bottom ash content has correlated to the slight increase in linear shrinkage. For example, with addition of 5% bottom ash into soil mixture, the LS value slightly increased by 3% from 21.2% to 24.2% and remained relatively unchanged at about 24% with further increase in bottom ash content. This could be attributed to the fact that the liquid limits of ash-soil mixtures decreased with the increase in the bottom ash content (Modarres and Nosoudy 2015). Since the same liquid limit was used for all ash-soil samples in linear shrinkage tests, an abundant amount of water over the level required for liquid limit was added into the samples as ash content increased from 5% to 30%. This brought about the shrinkage growth at the first mixture with 5% ash addition, potentially subjected to the evaporation of water from voids in the mixture. The higher ash-content samples or less proportions of soil in admixtures caused more free water flow outside the voids added. Their LS values, therefore, might not be affected by the increase in bottom ash content.

However, with addition of 5% hydrated lime in the blend of soil, the LS reduced significantly from 21.2% to about 15%. This result is derived from the introduction of hydrated lime, which is considered as chemical agent to activate the pozzolanic reaction in the mixture. Furthermore, the decreasing rate of LS was continuously steady when the lime-treated soil samples were supplemented with an increasing percentage of bottom ash from 5% to 10%, which caused shrinkage reduction from 13.2% to 11.3% (Fig. 1). This means that the combination of bottom ash and hydrated lime in soil yielded higher decrease in linear shrinkage than either lime-treated soil or ash-treated samples. With the addition of lime into soil-ash mixture, the LS results were in inverse correlation from negative outcome (increasing LS values) to positive one (decreasing LS values) and levelled off at about 8% linear shrinkage at the ash percentage of 20% and 30% (Fig. 1). This remarkable improvement of shrinkage can be attributable to hydrated lime addition, which causes the aggregation and flocculation of clay particles. This phenomenon reduces the clay surface area, inducing the soil mixture to be less clayey or plastic, and to become coarser. The characteristic of a coarser material like sand becomes gradually

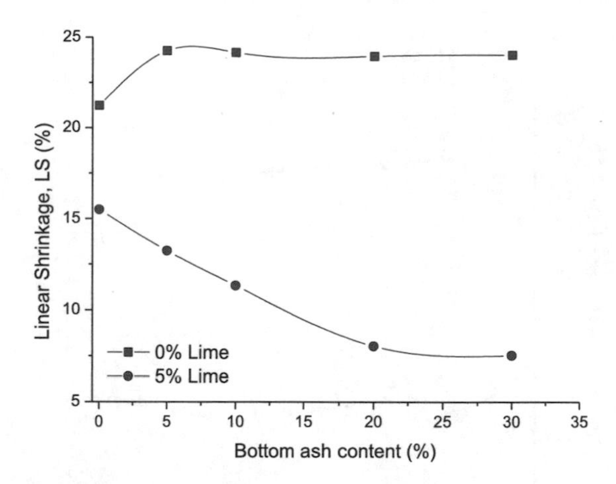

Fig. 1. Linear shrinkage of different contents of bottom ash treated expansive soil without or with 5% hydrated lime after 7 days of curing

overwhelming in the admixture to which more bottom ash is added; and as a result, the more bottom ash, the less linear shrinkage. Furthermore, this trend develops to the stage where the ash content is dominant in the blend, thus the shrinkage value remains unchanged. As observed in Fig. 1, for ash-lime treated soil, the linear shrinkage improvement when lime-ash combination exceeded 20% appeared to be minimal. In contrast, the blend of soil and bottom ash without lime combination indicated the constant high shrinkage at the first mix with only 5% ash. From this high level of shrinkage, the addition of 5% lime into ash-soil mixtures considerably reduced the linear shrinkage of soil-ash blend to around 8% when the ash content increased to 20% and then it remained constant with the further increase in the ash content up to 30%.

4.2 Unconfined Compression Strength

Figure 2 illustrates the effect of bottom ash content on unconfined compressive strength of treated expansive soil after 7 days of curing. An amount of 5% hydrated lime was also added to the ash-soil sample to examine the impacts of lime-ash combination on soil strength. As can be observed in Fig. 2, the UCS of ash-treated soil without lime fluctuated around 250 kPa, which was about 35 kPa lower than that of untreated soil. This implies that the addition of bottom ash into soils, after 7 curing days, slightly reduced the soil strength by 12.3% as the ash content increased from 5% to 30%. However, when 5% lime content was added into the ash-soil mixtures, the strength increased considerably to 552.5 kPa for soil treated with 5% bottom ash, which helped enhance the strength of ash-soil mixture by 325.6 kPa (approximately 143.5%). Therefore, it can be noted that bottom ash has an adverse effect on the soil strength for expansive soil modification if it is merely added to soil without supplementary reagents like lime or cement. This finding is in good agreement with previous studies on this kind of ash (Kamei et al. 2013, Lu et al. 2014, Modarres and Nosoudy 2015, Seco et al. 2011).

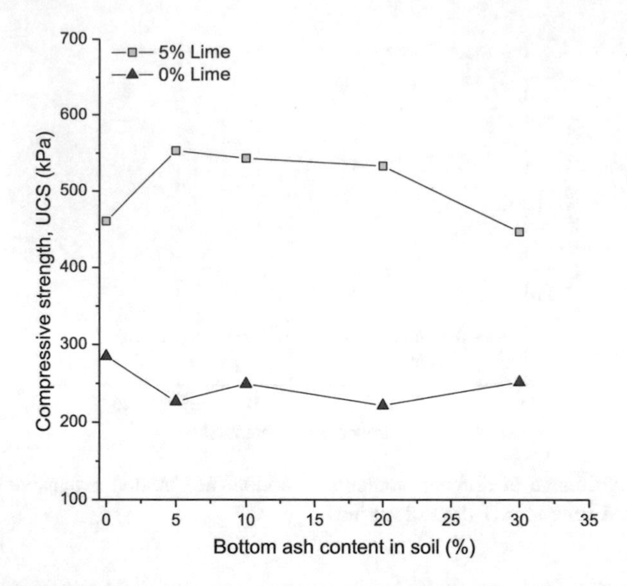

Fig. 2. Effect of bottom ash and hydrated lime on UCS of expansive soil after 7 days of curing

As can also be seen in Fig. 2, the 7-day-curing UCS values of lime-ash-treated soil showed a negligible decrease when the ash content approached 20%. Nevertheless, the strength still maintained a relatively same level at about 540 kPa before abruptly dropping to 446.8 kPa at the highest ash content (30%). This highlights the optimum combination content of lime and bottom ash, ranging from 5% to 20%, for expansive soil stabilisation. The optimum combination content is defined as the highest combined additive content to achieve the highest compressive strength value (about 500 kPa) while the use of bottom ash can be maximized in an effective way.

It is also worthwhile to note that the combined effect of hydrated lime and bottom ash proves their merit in the strength improvement of expansive soil. This enhancement could stem from pozzolanic reactions brought by combination of lime and silica from bottom ash. Consequently, cementitious bonds between clay particles developing in the moisturized lime-ash-soil mixtures hardens their structural strength. Their UCS, therefore, is higher than that of lime-treated soil. However, once the content of bottom ash is greater than a certain level (e.g. 30%) that ash particles are abundant and left from their pozzolanic reactions with lime, these free grains of bottom ash could weaken the bonds between clay elements. Hence, the reduction of compressive strength would be a consequence as what occurs when an excessive content of bottom ash is used (Fig. 2). The interpretation of this investigation is consistent with previous research conducted on the lime-ash combination to stabilise expansive soil (Gullu 2014; Ranga 2016).

Regarding the curing effect on the UCS values, a series of UCS tests was prepared and conducted after 7 and 28 days of curing. Figure 3 depicts the variation of compressive strength values for ash treated soil samples with or without lime combination obtained after curing for up to 28 days. It is clear from the figure that the higher UCS is proportional to the longer curing time for lime-ash-soil mixtures, but the strength remains stable in the case of soil-ash samples. For instance, when curing time increased from 7 to 28 days, there was about 32% increase in the UCS for lime-ash treated soil samples as ash content increasing from 0% to 10%. However, insignificant UCS increase is observed for soil-ash specimens. The strength of soil treated with only ash after 28 days of curing even indicated a mild decrease when the ash content increased from 5% to 30%. By way of illustration, the 28-day UCS value of soil stabilised with 30% bottom ash was lower than the corresponding 7-day one (236.6 kPa compared to 252.2 kPa, respectively). By contrast, even though lime-ash-soil UCS after curing for 28 days declined markedly when ash content increased from 20% to 30%, the 28-day soil strength at 30% bottom ash-lime combination was still higher than that of 7 day cured samples, showing a growth of 40% from 446.82 kPa to 625.89 kPa, respectively. This increase was double when compared with the approximately 20% improvement in UCS of only lime-soil mixture as curing time extended from 7 to 28 days. Significantly, with 20% content of combined bottom ash-lime treated soil, the 28 days compressive strength leaped to 820.68 kPa, which corresponded to a 54% improvement of UCS compared to the strength of soil treated with only lime at the same curing time. In order words, the strength improvement at 20% combined ash-lime content was more than 1.5 times, compared to the lime-soil UCS, and nearly triple the ash-soil UCS. Such high increase in the strength with prolonged curing time is probably attributable to reaction between lime and soil, leading to more cementitious linkage which is formed in soil aggregates. The strength in soil-ash blends, on the other hand, is not reinforced by this chemical reaction, and thus it does not show any improvement in UCS.

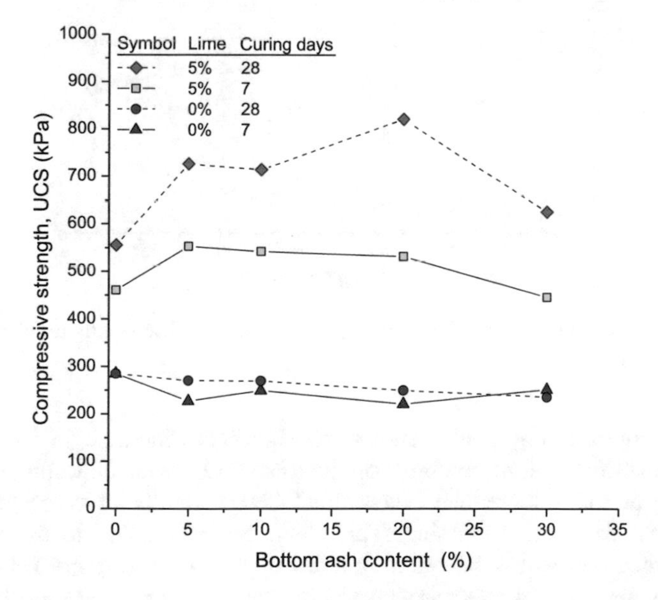

Fig. 3. Effect of bottom ash and lime on UCS of treated expansive soil at 7 and 28 days of curing

When it comes to longer curing days, the same pattern is repeated. Figures 4 and 5 illustrate the impact of longer curing time on UCS of bottom-ash- and hydrated-lime-bottom-ash-stabilized expansive soil, respectively. Overall, the UCS of soil-ash admixtures increased gradually with the longer curing time from 7 to 56 days (Fig. 4). However, this modest growth could not recover the initial UCS of parent soil samples which constitutes 285 kPa. The addition of 5% bottom ash into soil mixtures seemed to gain the same strength of untreated soil after curing for 56 days; meanwhile, the lower UCS value of about 260 kPa was observed for soils treated with higher bottom ash content at the same-curing-day, downgrading the original soil strength by 8.8%. Furthermore, whereas most specimens had upward trend of UCS after 28 days, the sample with 10% ash had a slight drop from the day of 28 to 56. Interestingly, as shown in Fig. 5, the similar phenomenon can be observed for soils treated with the lime-ash combination. After 28 days of curing, the compressive strength of 10% combined ash-lime treated soil suddenly went down to become the weakest sample at the curing day of 56. This low strength might be due to the binder dosage of bottom ash and its moisture content, which have a marked effect on the internal reaction of the admixture (Geetha and Ramamurthy 2011; Gullu 2014). Other reasons for this downward trend could be the physical properties of bottom ash related to particle size, surface properties, morphology and content of amorphous phases (Jaturapitakkul and Cheerarot 2003).

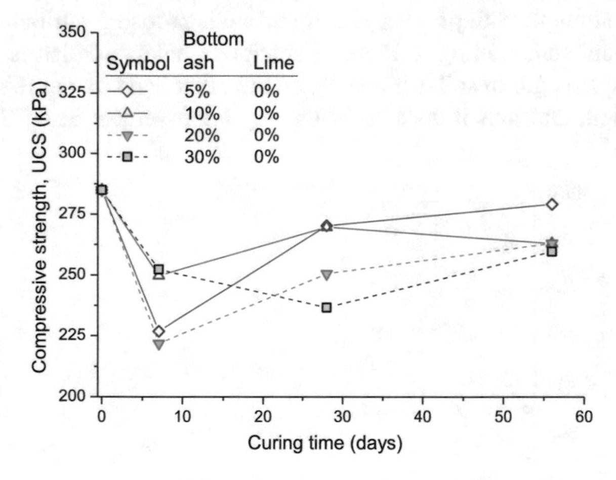

Fig. 4. Effect of curing time on UCS of expansive soil stabilised with different contents of bottom ash (0% Lime)

In stark contrast, Fig. 5 also shows the significant increase in UCS of treated expansive soil with ash-lime combination. It is also observed in all testing mixtures that after a curing period of more than 7 days, this UCS increase is set to accelerate at lower pace than that from the beginning. This could be attributable to the fact that the hardening process of soil stabilization in a few days after mixing creates cementitious bonds which burgeon and cover around soil particles. Such covering is a hindrance for allowing more lime penetrated and embedded to the soil aggregates. Consequently, the speed of forming the bonds decelerates and the strength development goes steadily.

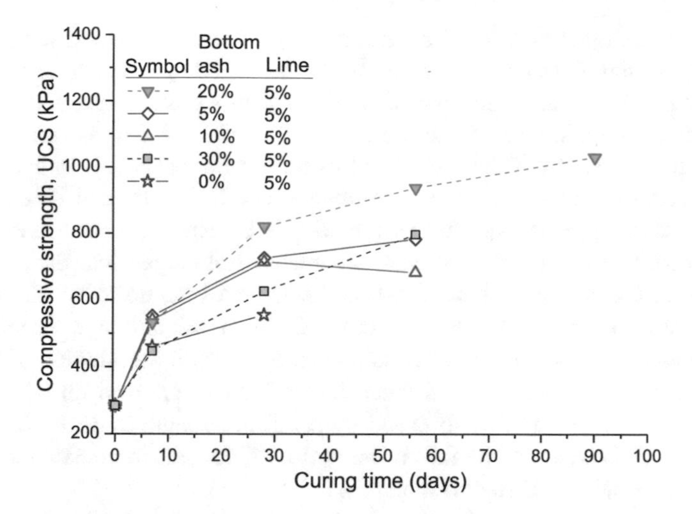

Fig. 5. Effect of curing time on UCS of 5% lime treated expansive soil with different contents of bottom ash

Moreover, the smaller increment of UCS strength after 7 days of curing may be the result of crystal development in the mixtures (Boardman et al. 2001).

In comparison with the strength of untreated expansive soil as presented in Fig. 5, the UCS increase of soil treated with 20% ash combined with 5% lime approximately doubled after 7 days, over tripled after 56 days and roughly quadrupled after 90 days of curing, becoming the highest strength sample all over the investigated curing days, and reaching the highest strength of 1 MPa. This indicates that the ratio of 20% bottom ash to 5% lime could be the optimum combination ratio of 1:4 for soil stabilisation. In addition, when compared with only lime treated soil, this combination ratio improved the strength of ash-lime-soil admixture by 15% after 7 days and 48% after 28 days of curing (Fig. 5). This finding indicates the combined effects of bottom ash and hydrated lime on expansive soil treatment. The combination yields the higher strength and the lower linear shrinkage than lime or bottom ash alone to stabilise soil. It is also noted that the utilization of bottom ash-hydrated lime combination for expansive soil treatment feasibly minimizes the adverse impact of the bottom ash (waste) on the environment. The application also provides cost-effective construction material via reduction of lime dosage and extra soil reinforcement by the combination of lime and bottom ash.

5 Conclusions

To examine the effect of bottom ash and its combination with hydrated lime on the shrinkage and the strength improvement of expansive soil, a series of experimental work were undertaken with several soil samples treated or untreated with these two stabilisers. From the findings of this investigation, the following conclusions can be drawn.

1. There was a slight increase in linear shrinkage of bottom ash-soil samples. On the contrary, a significant reduction of the linear shrinkage was observed for soils stabilised with hydrated lime and its addition with various contents of bottom ash. With the combination of 15% ash and 5% lime after 28-day of curing, the linear shrinkage reduced by 62.4% compared to that of untreated soil sample. With this combined ratio, adding bottom ash combined with lime exceeded 20% into the soil sample did not produce any further remarkable decrease in linear shrinkage.
2. The unconfined compressive strength of treated soil surged into the leap with the increase in the combined bottom ash and hydrated lime dosages. The increase in strength was apparently higher for combined bottom ash and lime treated soils than for bottom ash or hydrated lime alone treated soils. With the addition of bottom ash and lime combination, the UCS value levelled off or climbed up with increasing bottom ash content to 20%, followed by a fall in strength to 30%. This indicated that the 20% bottom ash might be the optimum content in combination with 5% lime for expansive soil stabilisation.
3. The strength development of hydrated lime and bottom ash stabilised expansive soil is also in proportional to the curing time increased. However, the rate of the strength development decelerated after 7 days of curing, which might result from the first reactions of pozzolan between lime and soil, forming cementitious links that hinders more lime reacting with clay particles. Furthermore, the strengths of 10% bottom ash treated soils without or with lime inclusion decreased when curing time extended from 28 days to 56 days. The possible reason for this phenomenon might be due to the change in moisture content, the binder dosage and the physical property of bottom ash.
4. The measured results of this experimental study demonstrated that the combined bottom ash and lime utilisation could be a promising solution for treatment of problematic soil because it can promote the higher strength and the lower linear shrinkage than only bottom ash or lime mixed with soil. Furthermore, utilisation of bottom ash to stabilise expansive soil helps to reduce the environmental consequence that the ash waste can cause. The cost-effective construction material by using bottom ash is another benefit when a certain dosage of lime for soil stabilisation can be reduced; this is because merely 5% lime combined with bottom ash in improving expansive soil can produce much higher compressive strength than that of parent soil.

Acknowledgments. This paper is part of an ongoing research at University of Technology Sydney (UTS) supported by Eraring Power Station, New South Wales, Arup Pty Ltd, Queensland Department of Transport and Main Roads (TMR), ARRB Group Ltd and Australian Sugar Milling Council (ASMC). The authors gratefully acknowledge their valuable assistance.

References

Vivi, A., et al.: Performance of chemically treated natural fibre and lime in soft soil for the utilisation as pile-supported earth platform. Int. J. Geosynthetics Ground Eng. (2015). https://doi.org/10.1007/s40891-015-0031-5

AS 1289.3.4.1: Methods of testing soils for engineering purposes - Soil classification tests - Determination of the linear shrinkage of a soil - Standard method. Standards Australia, Sydney, Australia (2008)

AS 1289.3.9.1: Methods of testing soils for engineering purposes – Soil classification tests - Determination of the cone liquid limit of a soil. Standards Australia, Sydney, Australia (2015)

ASTM D2487: Standard Practice for Classification of Soils for Engineering Purposes (Unified Soil Classification System). ASTM International, United States (2011)

Iman, B., Schoobbasti, C.J.: Stabilization of fine-grained soils by adding micro silica and lime or micro silica and cement. Electron. J. Geotech. Eng. **8**, 1–10 (2003)

Basha, E.M.A., et al.: Effect of the cement–rice husk ash on the plasticity and compaction of soil. Electron. J. Geotech. Eng. **8**, 1–8 (2003)

Bergado, D.T., et al.: Soft ground improvement in lowland and other environments. ASCE (1996)

Boardman, D.I., et al.: Development of stabilisation and solidification in lime–clay mixes. Geotechnique ICE (2001). https://doi.org/10.1680/geot.2001.51.6.533

Chen, F.H.: Foundations on Expansive Soils. Elsevier, Amsterdam (1988)

Cheriaf, M.J., et al.: Pozzolanic properties of pulverized coal combustion bottom ash. Cem. Concr. Res. (1999). https://doi.org/10.1016/S0008-8846(99)00098-8

Erdal, Ç.: Use of class C fly ashes for the stabilization of an expansive soil. J. Geotech. Geoenviron. Eng. ASCE (2001). https://doi.org/10.1061/(ASCE)1090-0241(2001)127:7(568)

Consoli, N.C., et al.: Influence of fiber and cement addition on behaviour of sandy soil. J. Geotech. Geoenviron. Eng. ASCE (1998). https://doi.org/10.1061/(ASCE)1090-0241(1998)124:12(1211)

Dang, L.C., et al.: Influence of bagasse ash and hydrated lime on strength and mechanical behaviour of stabilised expansive soil. In: GEOQuébec 2015 (2015)

Dang, L.C., et al.: Enhancing the engineering properties of expansive soil using bagasse ash and hydrated lime. Int. J. Geomate (2016). https://doi.org/10.21660/2016.25.5160

Dang, L.C., et al.: An experimental study on engineering behaviour of lime and bagasse fibre reinforced expansive soils. In: The 19th International Conference on Soil Mechanics and Geotechnical Engineering (19th ICSMGE) (2017)

Das, B.M., Sobhan, K.: Principles of Geotechnical Engineering, 8th edn. Cengage Learning, Stanmford (2013)

Edil, T.B., et al.: Stabilizing soft fine-grained soils with fly ash. J. Mater. Civ. Eng. ASCE (2006). https://doi.org/10.1061/(ASCE)0899-1561(2006)18:2(283)

Fatahi, B., et al.: Mechanical characteristics of soft clay treated with fibre and cement. Geosynthetics Int. ICE (2012). https://doi.org/10.1680/gein.12.00012

Fatahi, B., et al.: Shrinkage properties of soft clay treated with cement and geofibers. Geot. Geol. Eng. (2013). https://doi.org/10.1007/s10706-013-9666-y

Ganesan, K., et al.: Evaluation of bagasse ash as supplementary cementitious material. Cem. Concr. Compos. (2007). https://doi.org/10.1016/j.cemconcomp.2007.03.001

Geetha, S., Ramamurthy, K.: Properties of sintered low calcium bottom ash aggregate with clay binders. Constr. Build. Mater. (2011). https://doi.org/10.1016/j.conbuildmat.2010.11.051

Geliga, E.A., Ismail, D.S.A.: Geotechnical properties of fly ash and its application on soft soil stabilization. UNIMAS E J. Civ. Eng. **1**(2), 1–6 (2010)

Gourley, C.S., et al.: Expansive soils: TRL's research strategy. In: 1st International Symposium on Engineering Characteristics of Arid Soils, London (1993)

Goyal, A., et al.: Properties of sugarcane bagasse ash and its potential as cement - Pozzolana binder. In: Twelfth International Colloquium on Structural and Geotechnical Engineering, Cairo, Egypt (2007)

Gullu, H.: Function finding via genetic expression programming for strength and elastic properties of clay treated with bottom ash. Eng. Appl. Artif. Intell. (2014). https://doi.org/10.1016/j.engappai.2014.06.020

Han, J., Gabr, M.: Numerical analysis of geosynthetic-reinforced and pile-supported earth platforms over soft soil. J. Geotech. Geoenviron. Eng. ASCE (2002). https://doi.org/10.1061/(ASCE)1090-0241(2002)128:1(44)

Holtz, R.D., et al.: An Introduction to Geotechnical Engineering, 2nd edn. Pearson Education, USA (2011)

Huang, W.H., Lovell, C.: Bottom ash as embankment material. In: Geotechnics of Waste Fills—Theory and Practice. ASTM International (1990). https://doi.org/10.1520/STP25300S

Jones, L.D., Jefferson, I.: Expansive soils. In: ICE Manual of Geotechnical Engineering, pp. 413–441. ICE Publishing, London (2012)

Jorat, M.E., et al.: Engineering characteristics of kaolin mixed with various percentages of bottom ash. Electron. J. Geotech. Eng. **16**(2), 841–850 (2011)

Kamei, T., et al.: The use of recycled bassanite and coal ash to enhance the strength of very soft clay in dry and wet environmental conditions. Constr. Build. Mater. (2013). https://doi.org/10.1016/j.conbuildmat.2012.08.028

Kayabal, K., Buluş, G.: The usability of bottom ash as an engineering material when amended with different matrices. Eng. Geol. (2000). https://doi.org/10.1016/S0013-7952(99)00097-6

Khabbaz, H., Fatahi, B.: Stabilisation of closed landfill sites by fly ash using deep mixing method. In: Grouting and Deep Mixing 2012. ASCE (2012). https://doi.org/10.1061/9780784412350.0027

Kim, B.J., et al.: Geotechnical properties of fly and bottom ash mixtures for use in highway embankments. J. Geotech. Geoenviron. Eng. ASCE (2005). https://doi.org/10.1061/(ASCE)1090-0241(2005)131:7(914)

Kim, B., Prezzi, M.: Compaction characteristics and corrosivity of Indiana class-F fly and bottom ash mixtures. Constr. Build. Mater. (2008). https://doi.org/10.1016/j.conbuildmat.2006.09.007

Kim, Y.T., et al.: Experimental study on engineering characteristics of composite geomaterial for recycling dredged soil and bottom ash. Mar. Georesour. Geotechnol. (2011). https://doi.org/10.1080/1064119X.2010.514237

Kim, Y.T., Do, T.H.: Effect of bottom ash particle size on strength development in composite geomaterial. Eng. Geol. (2012). https://doi.org/10.1016/j.enggeo.2012.04.012

Kinuthia, J., Nidzam, R.: Effect of slag and siliceous additions on the performance of stabilized coal waste backfill. In: World of Coal Ash 2009 (WOCA) Conference, Lexington, KY, USA (2009)

Kolay, P.K., et al.: Tropical peat soil stabilization using Class F pond ash from coal fired power plant. World Acad. Sci. Eng. Technol. J. **2011**(74), 15–19 (2011)

Kunitomo, H.: Recycling of coal ash: current activities and challenges for the future. In: Proceedings of Symposium on Effective Use of Coal Ash, p. 23 (2009). (in Japanese)

Lav, A.H., Kenny, P.J.: The use of stabilised fly ash in pavements. In: National Symposium on the Use of Recycled Materials in Engineering Construction: 1996. Programme and Proceedings, Institution of Engineers, Australia, p. 77 (1996)

Le, T.M.: Analysing consolidation data to optimise elastic visco-plastic model parameters for soft clay. Ph.D. thesis, University of Technology Sydney, Sydney, Australia (2015)

Le, T.M., et al.: Numerical optimisation to obtain elastic viscoplastic model parameters for soft clay. Int. J. Plast. (2015). https://doi.org/10.1016/j.ijplas.2014.08.008
Liu, H.L., et al.: Performance of a geogrid-reinforced and pile-supported highway embankment over soft clay: case study. J. Geotech. Geoenviron. Eng. ASCE (2007). https://doi.org/10.1061/(ASCE)1090-0241(2007)133:12(1483)
Lorenzo, G.A., Bergado, D.T.: Fundamental parameters of cement-admixed clay—new approach. J. Geotech. Geoenviron. Eng. ASCE (2004). https://doi.org/10.1061/(ASCE)1090-0241(2004)130:10(1042)
Lu, S.G., et al.: Effect of rice husk biochar and coal fly ash on some physical properties of expansive clayey soil (Vertisol). Catena (2014). https://doi.org/10.1016/j.catena.2013.10.014
Mackiewicz, S.M., Ferguson, E.G.: Stabilization of soil with self-cementing coal ashes. In: Proceedings of 2005 World of Coal Ash (WOCA), Lexington, Kentucky, USA, 11–15 April 2005, p. 7 (2005)
Manikandan, A.T., Moganraj, M.: Consolidation and rebound characteristics of expansive soil by using lime and bagasse ash. Int. J. Res. Eng. Technol. (2014). https://doi.org/10.15623/ijret.2014.0304073
Miller, G.A., Azad, S.: Influence of soil type on stabilization with cement kiln dust. Constr. Build. Mater. (2000). https://doi.org/10.1016/S0950-0618(00)00007-6
Modarres, A., Nosoudy, Y.M.: Clay stabilization using coal waste and lime—technical and environmental impacts. Appl. Clay Sci. (2015). https://doi.org/10.1016/j.clay.2015.03.026
Nelson, J., Miller, D.J.: Expansive Soils: Problems and Practice in Foundation and Pavement Engineering. Wiley, New York (1997)
Nguyen, L.D., et al.: A constitutive model for cemented clays capturing cementation degradation. Int. J. Plast. (2014). https://doi.org/10.1016/j.ijplas.2014.01.007
Obuzor, G.N., et al.: Soil stabilization with lime activated GGBS—a mitigation to flooding effects on road structural layers/embankments constructed on floodplains. Eng. Geol. (2012). https://doi.org/10.1016/j.enggeo.2012.09.010
Osinubi, K.J., et al.: Bagasse ash stabilization of lateritic soil. In: Yanful, E. (ed.) Appropriate Technologies for Environmental Protection in the Developing World. Springer (2009). https://doi.org/10.1007/978-1-4020-9139-1_26
Phani Kumar, B.R.: A study of swelling characteristics and granular pile-anchor foundation technique in expansive soils. Ph.D. thesis, JN Technological University, Hyderabad, India (1997)
Phani Kumar, B.R.: Effect of lime and fly ash on swell, consolidation and shear strength characteristics of expansive clays_a comparative study. Geomech. Geoeng. Int. J. (2009). https://doi.org/10.1080/17486020902856983
Phani Kumar, B.R., Sharma, R.S.: Effect of fly ash on engineering properties of expansive soils. J. Geotech. Geoenviron. Eng. ASCE (2004). https://doi.org/10.1061/(ASCE)1090-0241(2004)130:7(764)
Rahman, M.D.A.: The potentials of some stabilizers for the use of lateritic soil in construction. Build. Environ. (1986). https://doi.org/10.1016/0360-1323(86)90008-9
Ranga, S.K.: Geotechnical characterisation of soft clay treated with a bottom and fly ash mixture. In: Geo-China 2016. ASCE (2016). https://doi.org/10.1061/9780784480014.015
Rifai, A., et al.: Characterization and effective utilization of coal ash as soil stabilization on road application. In: Geotechnical Society of Singapore - International Symposium on Ground Improvement Technologies and Case Histories (ISGI 2009), pp. 469–474 (2010). https://doi.org/10.3850/gi025
Satyanarayana, B.: Swelling pressure and related mechanical properties of block cotton soil. Ph.D. thesis, Indian Institute of Science, Bangalore (1996)

Seco, A., et al.: Stabilization of expansive soils for use in construction. Appl. Clay Sci. (2011). https://doi.org/10.1016/j.clay.2010.12.027

Sharma, R.S., et al.: Engineering behaviour of a remolded expansive clay blended with lime, calcium chloride, and rice-husk ash. J. Mater. Civ. Eng. ASCE (2008). https://doi.org/10.1061/(ASCE)0899-1561(2008)20:8(509)

Shibi, T., Kamei, T.: Effect of freeze–thaw cycles on the strength and physical properties of cement stabilized soil containing recycled bassanite and coal ash. Cold Reg. Sci. Technol. (2014). https://doi.org/10.1016/j.coldregions.2014.06.005

Sivakumar Babu, G.L., et al.: Use of coir fibers for improving the engineering properties of expansive soils. J. Nat. Fibers (2008). https://doi.org/10.1080/15440470801901522

Viswanadham, B.V.S., et al.: Swelling behaviour of a geofiber-reinforced expansive soil. Geotext. Geomembr. (2009). https://doi.org/10.1016/j.geotexmem.2008.06.002

Enhancement of Expansive Soil Properties Using Cement Kiln Dust Mixed with Lime

Qassun S. Mohammed Shafiqu(✉) and Reem I. Abass

Civil Engineering Department, Al-Nahrain University, Baghdad, Iraq
qassun@yahoo.com

Abstract. The behavior of expansive clays in the form of shrinkage and swelling caused by changes in moisture content is frequently expressed superficially as settlement and heaving of lightly loaded geotechnical structures such as roadways, pavements, railways, foundations and channels or linings of reservoir. Research studies has been carried out previously in various parts of the world to find out economical and efficient ways and means of using cement kiln dust (CD) and lime (L) individually in various applications like soil stabilization. This research dealt with stabilization of an expansive high plastic soil using cement kiln dust (CD) with lime (L) to reduce their swelling and improve their geotechnical properties. The swelling potential is 1% measured from free swell odometer test. In this study, the tests results such as compaction, Atterberg limits, swelling potential, California Bearing Ratio (CBR) and unconfined compressive strength obtained on expansive clays mixed at different proportions of cement kiln dust-lime are presented and discussed. The results indicates that swelling potential decreased from 19% to 2% by adding (16% CD + 6% L) and the highest value for CBR is obtained by adding (16% CD + 9% L) as these percentages improved the CBR from 5.45 to 35.95.

1 Introduction

Expansive clayey soils are not only possess the tendency to swell or increase in volume but also to shrink or decrease in volume when the prevailing moisture content is allowed to change. Such change of water content of these soils can take place from rains, floods, or leakage of sewer lines. The response of expansive soils in the form of swelling and shrinkage due to changes in water content is frequently expressed superficially as heaving and settlement of lightly loaded geotechnical structures. Marl and silty mudstone, bentonitic mudstone, argillaceous lime-stone and altered conglomerate are considered as expansive materials that exhibit swelling problems. Consequently, expansive soils cause distress and damage to structures founded on them (Amer 2006). Stabilizing soil to have required engineering specifications is carried out using many procedures. The methods range from mechanical to chemical stabilization. Many of these methods can be considered expensive to be implemented by slowly developing countries and the best way is to use local materials which are available and relatively cheap costs affordable by their internal funds (Magafu and Li 2010).

In this study, a chemical method for treating the swelling of prepared expansive clayey soil is presented using cement kiln dust-lime. The full experimental program is

J. S. McCartney and L. R. Hoyos (Eds.): GeoMEast 2018, SUCI, pp. 45–55, 2019.
https://doi.org/10.1007/978-3-030-01914-3_4

carried out for studying adding effect of cement kiln dust-lime with different percentages by weight on the consistency limits, compaction characteristics, unconfined compressive strength with curing, swelling and swelling pressure and California Bearing Ratio values.

2 Materials and Methods

2.1 Natural Soil

The soil samples are collected from a depth of (2–2.5 m) below the soil surface from Al-Karada at Baghdad city. The soil used in this study was silty clay (62% clay + 38% silt). This natural soil is subjected to laboratory tests at laboratory of civil engineering in university of Al-Nahrain to determine its physical properties. The classification of the natural soil was CL according to the USCS.

2.2 Bentonites

The Bentonite used to make artificial expensive soil is manufactured by Turkey company limited.

2.3 Prepared Soil

The expansive clayey soil was prepared in laboratory by mixing (40%) of bentonite with (60%) of natural soil (Kaolinite) in order to achieved the research requirements to reduce the swell potential of expansive soils by adding cement kiln dust-lime with different percentages and studied their effects. The swell of prepared soil reached to (19%).

2.4 Cement Kiln Dust (CKD)

Cement kiln dust used in this study was brought from Al-Kufa plant. Treatment with cement dust was found to be an effective option for improvement of soil properties, and easy to use and mix with soil due to it's fine particle size (Harichane et al. 2010). Cement dust is used to improve properties of expansive soil.

2.5 Lime (L)

The lime used in this study was Quicklime produced in Iran.

3 Preperation of Test Sample

The expansive soil was constituted artificially in laboratory by mixing of natural soil with bentonite. The artificial expansive soil, which was stated in this study, comprised of 60% of natural soil and 40% of bentonite by dry mass. At first, natural soil was placed inside oven at 105 °C for 24 h to ensure complete dryness. The pulverization step followed the drying. Pulverization was provided by Los Angeles machine. Then, cement kiln dust-lime materials used in this study were mixed with prepared soil by mixer until the color of mixture becomes homogenous and then put into nylon bags.

4 Experimental Work

An experimental program was planned and implemented for the sake of investigating the improvement of the expansive soil. Atterberg limit (ASTM D4318.00), compaction test (ASTM D698-00), swell and swell pressure tests (ASTM D4546-14), unconfined compression test (ASTM D2166-00) and California Bearing Ratio (CBR) test (ASTM D1883-87) are conducted on expansive soil and stabilized soil. These tests were carried out in soil mechanics laboratory, College of Engineering in Al-Nahrain university. Chemical tests on prepared expansive soil, cement kiln dust-lime were carried out in the physics laboratory, College of Science in Al-Nahrain University. The physical and chemical properties for prepared expansive soil as well as chemical composition of cement kiln dust-lime are shown in Tables 1, 2, 3 and 4 respectively.

Table 1. Summary of physical and classification tests for prepared expansive soil

Properties	Standard	Expansive soil
Sand (%)	ASTM D 422	0
Silt (%)	ASTM D 422	20
Clay (%)	ASTM D 422	80
Liquid limit	ASTM D4318	95
Plastic limit	ASTM D4318	27
Plasticity index	ASTM D4318	68
Optimum moisture content (%)	ASTM D 698	28
Max. dry density (kN/m^3)	ASTM D 698	1.47
Specific gravity, Gs	ASTM D 854	2.65

Table 2. Chemical tests results for prepared expansive clayey soil

SiO_2%	Al_2O_3%	CaO%	MgO%	FeO%	Na_2O%	SO_3%	BaO%	Sn%
64.18	14.71	8.00	4.30	2.43	2.28	1.37	0.82	1.90

Table 3. Chemical composition cement kiln dust

CaO%	SiO_2%	SO_3%	Al_2O_3%	MgO%	K_2O%	Na_2O%	FeO%	Sn%	Cl%
44.61	17.48	8.12	5.22	4.29	4.23	1.73	1.02	12.18	1.13

Table 4. Chemical composition of lime

CaO%	MgO%	Al_2O_3%	SiO_2%	SO_3%	Sn%
71.36	4.15	3.00	2.56	1.67	17.26

5 Results

The following are the results of the experimental tests carried out in this study:

5.1 Atterberg's Limits

Atterberg limits (liquid and plastic limits) according to (ASTM D4318.00). This test is performed to determine the plastic and liquid limits of fine grain soil that pass the sieve (0.425 mm). The effects of adding different percentages of (CKD & L) on the consistency limits values of the prepared expansive clayey soil are summarized in Table 5. In general, liquid limit (LL) and plasticity index (PI) decrease with increasing amount of (CKD & L), while plastic limit (PL) increases with increasing (CKD & L). The maximum decrease in liquid limit is found with the addition of (6% Lime + 16% CKD), this addition reduced the liquid limit by about 28.5%. The maximum increase in plastic limit is found with the addition of (9% L + 12% CKD) this addition increased the plastic limit to 34%. The maximum decrease in plasticity index is found with the addition of (6%Lime + 16%CKD) this addition reduced the plasticity index by about 47%. Figure 1 shows the adding influence of (CKD & L) on L.L.%, P.L. % and P.I.% of expansive clayey soil. The addition of cement kiln dust-lime led to a reduction of L.L.%, P.I.% and increase in P.L.%. This reduction was due to a decrease in the thickness of the double layer of the clay particles. That is because of cation exchange reaction, which causes an increase in the attraction force leading to a flocculation of the particles (Nalbantoğlu 2004).

Table 5. Results of expansive soil Atterberg limits

Sample no.	CKD %	Lime %	L.L. %	P.L. %	P.I. %	Sample no.	CKD %	Lime %	L.L. %	P.L. %	P.I. %
1	0	0	95	27	68	9	12	0	76	29	47
2	0	3	86	28	58	10	12	3	73	30	43
3	0	6	76	29	47	11	12	6	71	31	40
4	0	9	74	31	43	12	12	9	71	34	37
5	8	0	89	28	61	13	16	0	73	30	43
6	8	3	78	29	49	14	16	3	72	31	41
7	8	6	73	30	43	15	16	6	68	32	36
8	8	9	73	32	41	16	16	9	70	33	37

5.2 Compaction Tests

This test used to determine the relationships between dry unit weight and water content of clay soil (compaction curve) according to (ASTM D698-00). The adding influence for different percentages of (CKD &L) on maximum dry density (MDD) and optimum

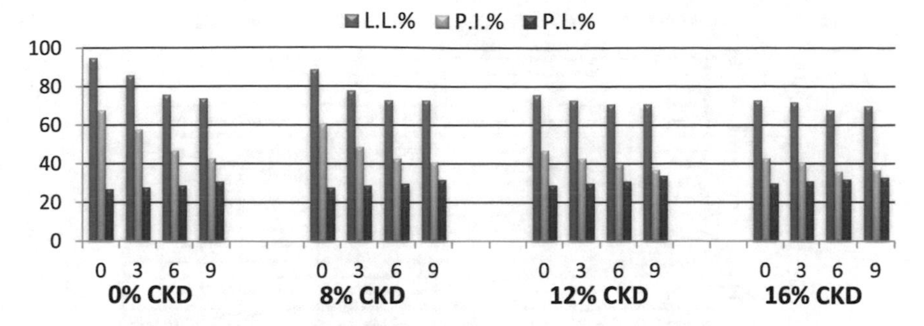

Fig. 1. Adding effect of (CKD & L) on L.L.%, P.L. % and P.I.%

moisture content (OMC) of expansive clayey soil is summarized in Table 6. Figure 2 shows the values of maximum dry density (MDD) at different percentages of (CKD &L) addition, for all percentages of (CKD & L) the maximum dry density increases with the increase in adding percent. Figure 3 shows the adding effect for different percentages of (CKD & L) on the optimum moisture content (OMC). It has been indicated that the optimum moisture content decrease with increasing percentage of adding (CKD & L). The decrease in the optimum water content is due to the change in surface area of composite samples. The increase in the maximum dry density depending on the behavior of clayey soil when it is mixed with (CKD & L) mixture which may change its texture from plastic to non-plastic soils and this phenomenon increases with the increase of adding (CKD & L) mixture.

Table 6. Adding effect of (CKD & L) on maximum dry density (MDD) and optimum moisture content (OMC)

Sample no.	CKD %	Lime %	MDD (gm/cm3)	OMC %	Sample no.	CKD %	Lime %	MDD (gm/cm3)	OMC %
1	0	0	1.470	28	9	12	0	1.498	22
2	0	3	1.460	26	10	12	3	1.500	24
3	0	6	1.442	25	11	12	6	1.502	23
4	0	9	1.430	24	12	12	9	1.512	22
5	8	0	1.485	23	13	16	0	1.518	21
6	8	3	1.488	25	14	16	3	1.520	23
7	8	6	1.490	24	15	16	6	1.522	22
8	8	9	1.495	23	16	16	9	1.550	21

5.3 Free Swell and Swell Pressure

Generally, the laboratory test method to measure the magnitude of one-dimensional wetting-induced free swell of unsaturated compacted soils are conducted by simple odometer test apparatus according to (ASTM D4546-14). The swelling pressure is the maximum external load which should be exerted to the soil to prevent expansive soil from any more deformation while wetting. The test is completed and the total pressure

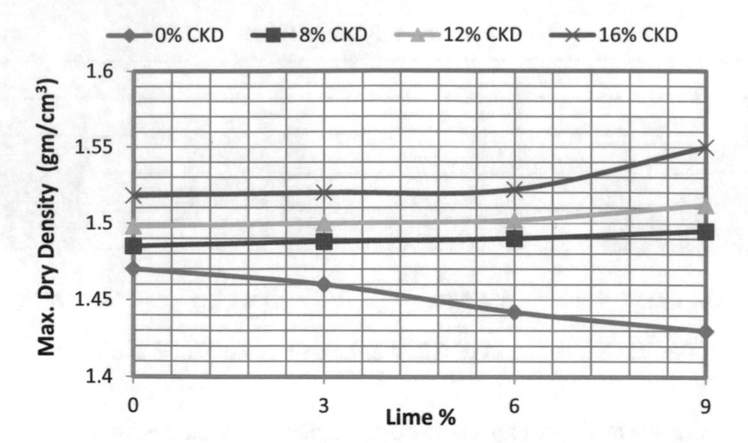

Fig. 2. Adding effect of (CKD & L) on MDD

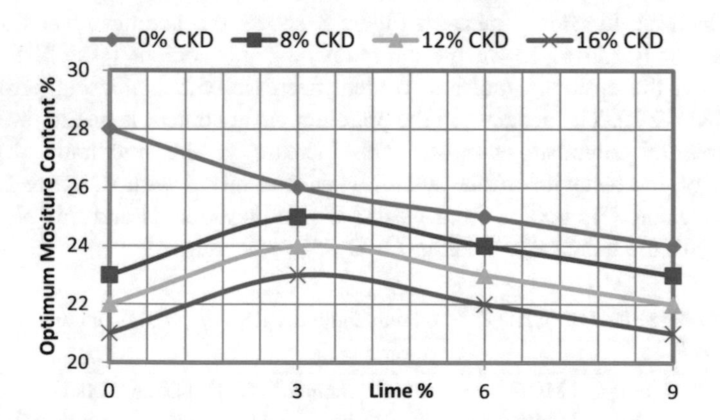

Fig. 3. Adding effect of (CKD & L) on OMC

that applied to prevent soil sample expansion is referred as swelling pressure. The addition of different percentage of (CKD & L) decreased the swelling of prepared soil mixtures. The free swell decreased from 19% to 2% by adding (16% CKD + 6% L). The adding effect for different percentage of (CKD & L) on the free swell% and swell pressure are shown in Table 7. Figures 4 and 5 shows the adding effect of (CKD & L) on free swell and swelling pressure of expansive soil respectively. In general, the swelling pressure values decrease with increasing (CKD & L) percentage. With the addition of (16% CKD + 6% L) the swelling pressure decreased from an initial value of (199.85) kPa for the prepared expansive clayey soil to (88.6) kPa.

5.4 Unconfined Compression Test (UCT)

The unconfined compression test achieved according to (ASTM D2166-00). The unconfined compressive strength is determined by applying an axial stress to cylindrical soil specimen with no confining pressure and observing the axial strains

Table 7. Adding effect of (CKD & L) on free swell

Sample no.	CKD %	Lime %	Free swell S %	Swell pressure (kPa)	Sample no.	CKD %	Lime %	Free swell S %	Swell pressure (kPa)
1	0	0	19	199.85	9	12	0	7	155.39
2	0	3	13	180.16	10	12	3	7	127.33
3	0	6	9	174.52	11	12	6	6	100.98
4	0	9	7	150.99	12	12	9	5	91.57
5	8	0	8	169.68	13	16	0	5	147.42
6	8	3	9	150.44	14	16	3	5	104.39
7	8	6	7	117.98	15	16	6	2	88.6
8	8	9	6	108.02	16	16	9	3	74.93

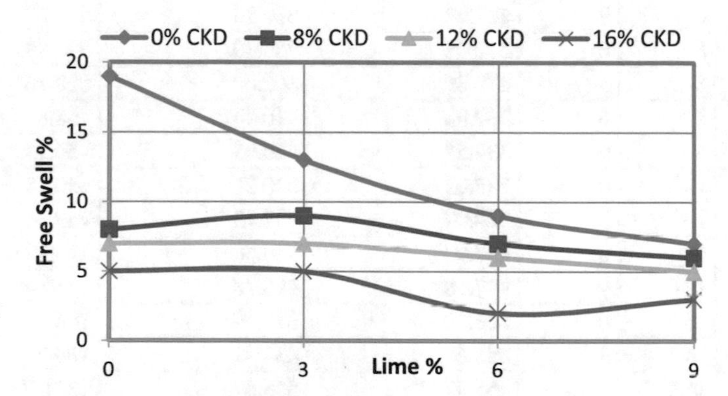

Fig. 4. Adding effect of (CKD & L) on free swell %

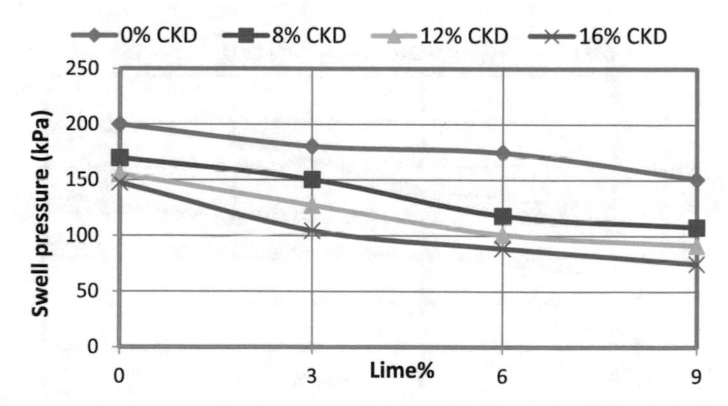

Fig. 5. Adding effect of (CKD & L) on swelling pressure

corresponding to various stress levels. The effects of cement dust, lime and polymethacrylate content, and curing periods (7 and 28 days) on the unconfined compressive strength were investigated. The increase in (CKD &L) percentage and curing period has improved the unconfined compressive strength. The maximum percent for

(CKD&L) that achieves the maximum strength is at (16% CKD + 9% L) for both 7 days and 28 days curing which shows increasing in the UCS value up to 574.51 kPa and 1598.65 kPa for 7 days and 28 days curing period respectively. Table 8 and Figs. 6, 7 and 8 shows the adding effect of (CKD &L) on expansive soils.

Table 8. Adding effect of (CKD & L) on Unconfined Compressive Strength (UCS)

Sample no.	CKD %	Lime %	UCS without curing (kPa)	UCS 7 days curing (kPa)	UCS 28 days curing (kPa)
1	0	0	160.98	–	–
2	0	3	204.77	280.33	450.87
3	0	6	217.07	330.76	483.67
4	0	9	227.42	389.86	560.67
5	8	0	270.02	286.80	578.89
6	8	3	279.89	302.60	628.76
7	8	6	287.02	410.44	768.89
8	8	9	337.55	420.65	956.12
9	12	0	287.02	310.21	650.93
10	12	3	320.02	342.43	789.34
11	12	6	487.97	560.66	1360.87
12	12	9	457.48	493.78	1210.45
13	16	0	306.19	375.40	746.7
14	16	3	353.66	433.87	984.21
15	16	6	455.97	548.87	1321.65
16	16	9	488.49	574.51	1598.65

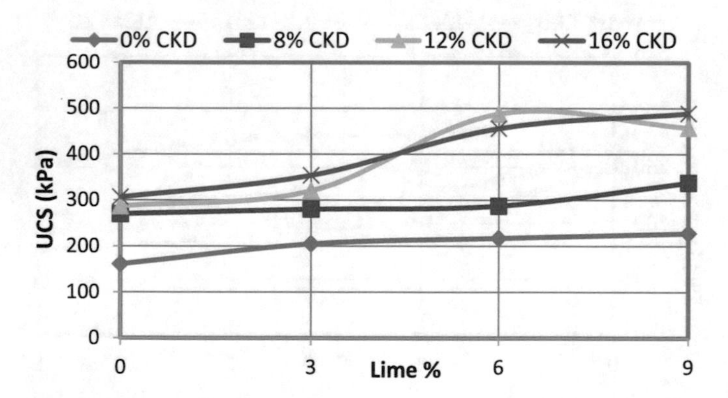

Fig. 6. Adding effect of (CKD & L) percentage on UCS without curing

5.5 California Bearing Ratio (CBR)

The CBR (California Bearing Ratio) is used to evaluate the potential strength of treated and untreated soils from laboratory compacted specimens according to (ASTM D1883-87). Table 9 and Fig. 9 show the adding influence of (CKD & L) on California Bearing Ratio

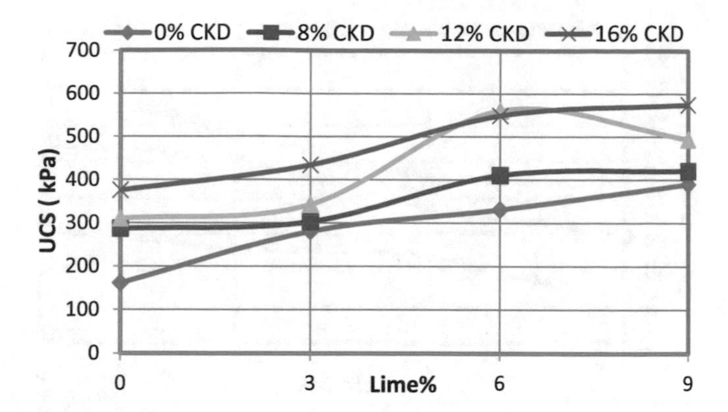

Fig. 7. Adding effect of (CKD & L) on UCS for 7 days curing period

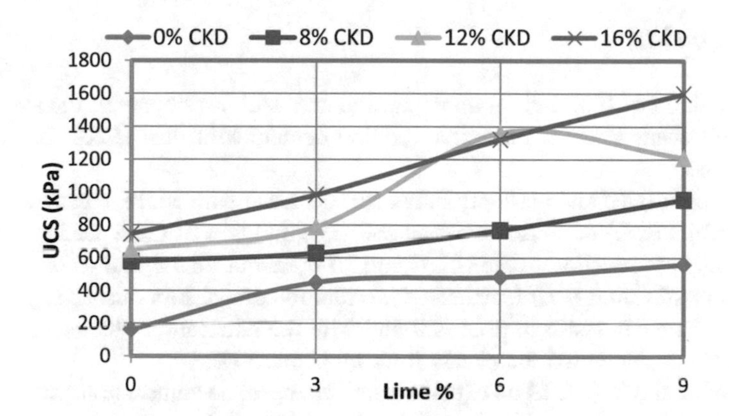

Fig. 8. Adding effect of (CKD & L) on UCS for 28 days curing period

Table 9. Adding effect of (CKD & L) on CBR

CBR		Lime %			
		0%	3%	6%	9%
CKD%	0%	5.45	11.68	13.04	15.99
	8%	14.74	18.71	25.17	27.62
	12%	19.89	20.34	29.50	28.99
	16%	21.53	22.93	30.01	35.95

(CBR) values of the expansive clayey soil. In general CBR increases with increasing (CKD & L) and the results show that the addition of cement kiln dust alone is more effective than adding lime alone on CBR values. The maximum percentage of (CKD & L) gives the highest value for CBR is by adding (16% CKD + 9% L) as this percentage improved the CBR value from 5.45 to 32.95. Also adding (16% CKD + 6% L) improves the CBR to 30.01.

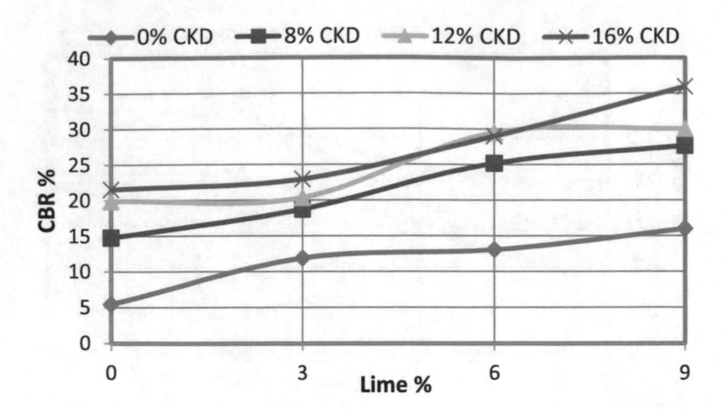

Fig. 9. Adding effect of (CKD & L) on California bearing ratio (CBR)

6 Conclusion

1. The addition of (CKD & L) on expansive clay soil has improved the swelling and strength characteristics more than adding cement kiln dust (CKD & L) or lime (L) alone.
2. Liquid limit (LL) and plasticity index (PI) decrease with addition cement kiln dust-lime (CKD & L) on expansive clay soil. The addition of (16% CKD + 6% Lime) gives higher reduction in the (LL) and (PI) by about 28.5% and 47% respectively. While plastic limit (PL) increase with addition cement kiln dust-lime (CKD & L) and the larger increase in (PL) is found with the addition of (12% CKD + 9% L), this addition increased the plastic limit by about 34%.
3. The addition (CKD & L) on expansive soil changes the compaction parameters. The maximum dry density increase with increase (CKD & L) content from 1.47 gm/cm^3 to 1.55 gm/cm^3 while optimum moisture content decreases from 28% to 21% with increase (CKD & L) content.
4. The addition of different percentage of (CKD & L) decrease the swelling potential and swell pressure of prepared soil mixtures. The free swell reduced from 19% to 2% by adding (16% CKD + 6% L) and swelling pressure decreased from an initial value of (199.85) kPa for the prepared expansive clayey soil to (74.93) kPa at (16% CKD + 9% L).
5. The increase in (CKD & L) percentage and curing period has improved the unconfined compressive strength. The maximum percent for (CKD & L) that achieves the maximum strength is at (16% CKD + 9%L) for both 7 days and 28 days curing which shows increasing in the (UCS) value by about 2.57 and 8.93 times the original values of UCS for 7 days and 28 days curing period respectively.
6. The (CBR) increases with increasing (CKD & L) content, the maximum percentage of (CKD & L) (i.e., 16% CKD + 9% L) gives the highest improvement in the value of (CBR) from 5.45 to 35.95 with a percentage of enhancement about 5.6 times the original values of CBR.

References

Amer, A.A., Mattheus, F.A.G.: Expansive soils recent advances in characterization and treatment. In: Proceedings and Monographs in Engineering, Water and Earth Sciences, pp. 1–6 (2006)

Magafu, F., Li, W.: Utilization of local available materials to stabilize native soil (earth roads) in Tanzania. case study Ngaral Engineering, vol. 2, pp 516–519 (2010)

Harichane, K., Ghrici, M., Khebizi, W., Missoum, H.: Effect of the combination of lime and natural pozzolana on the durability of clayey soils. Electron. J. Geotech. Eng. **15**(L), 1194–1210 (2010)

ASTM D4318.00. Standard Test Method for liquid limit, plastic limit, and plasticity index of soil. Annual Book of ASTM Standards, vol. 04(08)

Nalbantoğlu, Z.: Effectiveness of class C fly ash as an expansive soil stabilizer. Constr. Build. Mater. 377–381 (2004)

ASTM D698-00. Standard Test Methods for Laboratory Compaction Characteristics Using Standard Effort (600 kN-m/m3)

ASTM D 4546-14. Standard Test Method for One Dimensional Swell or Collapse of Soils

ASTM D2166-00. Standard Test Method for Unconfined Compressive Strength of Cohesive Soil

ASTM D1883-87. Standard Test Method for CBR (California Bearing Ratio) of Laboratory-Compacted Soils

Effect of Chemical Composition of Woodash and Lime on Stabilization of Expansive Soil

Tochukwu A. S. Ugwoke(✉) and Chukwuebuka Emeh

Department of Geology, University of Nigeria, Nsukka, Nigeria
tcugwoke@yahoo.comm

Abstract. Stabilizing effect of combined woodash and lime on expansive soil from south-eastern Nigeria has been evaluated. The evaluation followed subjecting industrial woodash to X-ray fluorescence (XRF) to determine its chemical composition while an expansive soil underlain by the Coniacian Awgu Group was subjected to X-ray diffraction (XRD) to determine the mineralogy of the soil. The plasticity index (PI), linear shrinkage (LS), free swell index (FSI) and unconfined compressive strength (UCS) of the soil was determined to ascertain the geotechnical properties of the natural soil. The soil was then mixed with woodash in varying proportions viz: 0% woodash and 100% soil; 6% woodash and 94% soil; 12% woodash and 88% soil; 18% woodash and 82% soil; 24% woodash and 76% soil. The PI, LS, FSI and UCS of each woodash-soil admixture was determined to ascertain how these geotechnical properties varies amongst the admixtures and thus the soil improvement of the various woodash proportions. The woodash-soil admixture that gave the best improvement quality was further mixed with 2%, 4%, 6% and 8% lime and PI, LS, FSI and UCS of each woodash-soil-lime admixture also determined to ascertain the amount of lime that gives the best improvement to the woodash-soil admixture. The XRF result revealed that the woodash was dominated with CaO and some other oxides in minor quantities. The XRD result revealed that the soil contains clay minerals. The geotechnical properties of the woodash-soil admixtures indicate that 18% woodash and 82% soil showed the best improvement in PI, SL, FSI and UCS of the soil while the addition of 4% lime to this best improved woodash-soil admixture further improved only the FSI and UCS. Results show that the stabilizing effect of both the woodash and combined woodash and lime is controlled by the chemical composition of the woodash.

Keywords: Woodash · Lime · Expansive soil · Stabilization
Chemical complexity

1 Introduction

Soil stabilization can be achieved either mechanically or chemically. Conventional chemical stabilizers like lime and Portland cement are more effective than mechanical stabilization because the high calcium oxide (CaO) content of these chemicals undergo cation exchange, flocculation and time dependent pozzalanic reaction with clay minerals contained in expansive soil while mechanical stabilization only reduces the void ratio of soil (Bell 1989; Muntohar and Hantoro 2000; and Show et al. 2003).

J. S. McCartney and L. R. Hoyos (Eds.): GeoMEast 2018, SUCI, pp. 56–63, 2019.
https://doi.org/10.1007/978-3-030-01914-3_5

Researchers had shown that some industrial waste materials like marble dust and limestone waste dust can also serve as soil stabilizers due to their high CaO content. Okagbue and Onyeobi (1999) and Agrawal and Gupta (2011) discovered that the addition of 8% marble dust to 92% soil increased the soil strength. Brook (2009) used rice husk and got an increased strength and reduction in swelling ability while Ji-ru and Xing (2002) used fly-ash (by-production of coal power plant) and got results similar to the above stated. Some researchers had combined the conventional and unconventional chemical stabilizers with a view of boosting their stabilizing potentials. Work by Rao et al. (2012) revealed that the addition of 25% rice husk and 9% lime to marine clay improved the plasticity index, California bearing ratio and differential free swell of the clay. Ismaiel (2006) and Malhotra and Naval (2013) combined fly ash and lime in the stabilization of soil and realized that the combined stabilizers (fly ash and lime) have more stabilization effect on the soil than either fly ash or lime. The authors did not, however, evaluate the extent to which the stabilizers' chemistry contributes to their stabilization effect.

This work discusses the effect of woodash chemical complexity on the stabilization effect of combined woodash and lime on soil. Findings in this work shall be helpful in ascertaining the cause of varying behaviours of woodash in stabilization of expansive soil.

2 Sampling and Laboratory Tests

About 500 g of lime (calcium oxide) used for this study was obtained from a chemical shop. The bottle containing the lime bears the chemical composition of the lime. About 3 kg of woodash, collected from Ife-best bakery (south-eastern Nigeria), was removed from the furnace, left undisturbed for 24 h to cool and passed through British Standard (BS) sieve 240 (63 μm) to obtain the size needed for ash-clay reaction. About 5 g of the woodash was subjected to x-ray fluorescence to determine its chemical composition while the remaining preserved in an airtight bag. The soil used was collected from Awgu town of south-eastern Nigeria which is characterised by cracked buildings and failed roads (Fig. 1a-d). Agwu town is underlain by the Coniacian Agwu shale, which tectonic origin and lithostratigraphy had been discussed by Murat (1972), Olade (1975), Ojoh (1990), Abubakar and Obaje (2001). About 5 kg bulk soil was collected from the area that showed the worst structural damage at a depth in excess of 35 cm. The soil was air-dried and divided into 6 portions.

The first portion of the soil was subjected to x-ray diffraction and quantified following methods described in Okogbue and Ugwoke (2015). The second portion was subjected to Atterberg limit tests, linear shrinkage, free swell and unconfined compressive strength (UCS) tests following BS 1377-2 (1990), ASTM C356 (2010), ASTM D4546 (2014) and ASTM D2166/D2166M (2016) standards respectively. The Atterberg limits were used in calculating the plasticity index. The remaining four soil sample portions were each thoroughly mixed with woodash in the following proportions (by mass): 94%, 88%, 82% and 76% soil was mixed with 6%, 12%, 18% and 24% woodash respectively. Each of the soil-woodash admixtures was subjected to the Atterberg limits, linear shrinkage, free swell and UCS tests following the standards

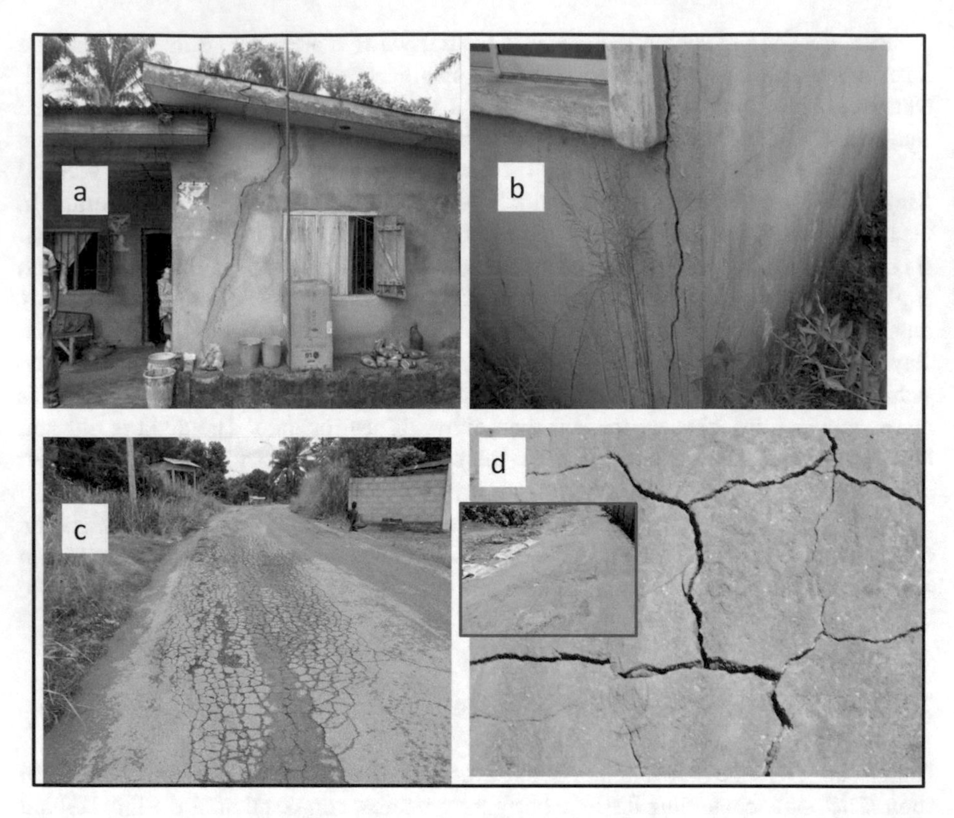

Fig. 1. Some failed civil engineering structures observed at Awgu town, Enugu State, south-eastern Nigeria (a and b-cracked buildings, c-failed tarred road, d-cracked untarred road)

earlier stated. The soil-woodash admixture (SWA) that gave the best improvement quality was further mixed with lime in the following proportions (by mass): 98%, 96%, 94% and 92% SWA was mixed with 2%, 4%, 6% and 8% lime respectively; each SWA-lime admixture was also subjected to the tests following standards earlier stated. The mixing processes were each done with the aid of hand trowel following quartering method on clean paved surfaces to avoid contamination.

3 Results and Discussion

The properties of the lime and woodash are shown in Table 1 while properties of the natural soil are shown in Table 2. It can be seen that, as expected, the dominant chemical composition in both lime and woodash is CaO. The CaO content of lime is more than that of woodash. Both lime and woodash have stabilizing potential but lime has higher stabilizing potential than woodash. Woodash contains many other oxides like P_2O_5, K_2O, MnO, Fe_2O_3 and Ag_2O in minor quantities. This chemical complexity of woodash (unlike lime), understandably, is because the proto-woodash (natural wood) has natural origin while lime was industrially manufactured. As shown in

Table 2, the natural soil is classified as inorganic clayey soil of high plasticity (CH) which renders it naturally unsuitable as sub-grade, sub-base, base or foundation material. The plasticity index, swell potential and clay mineralogy (>60%) of the soil also agree with its expansive nature.

Table 1. Properties of the woodash and lime

	Stabilizers	
Compounds/Property	Woodash (%)	Lime (%)
P_2O_5	3.40	–
SO_3	1.82	–
K_2O	15.1	–
CaO	71.58	96.00
TiO_2	0.46	–
Cr_2O_3	0.02	–
V_2O_5	0.09	–
MnO	2.37	–
MgO	–	0.80
Fe_2O_3	2.30	0.10
CuO	0.07	–
ZnO	0.17	–
Ag_2O	2.10	–
BaO	0.40	–
Re_2O_7	0.2	–
Loss On Ignition (LOI)	0.01	0.10
pH	12–13	13–14
Specific gravity	2.81	–

The effect of varying proportions of the woodash on plasticity index (PI), linear shrinkage (LS) and free swell index (FSI) of the soil is shown in Fig. 2. As the lowest PI, LS and FSI was attain on the addition of 18% woodash to the soil, this admixture (18% woodash and 82% soil) was taken as the optimum soil-woodash admixture (OSWA) and subsequently mixed with varying proportions of lime as explained earlier. Effects of the lime-OSWA admixture on the above stated properties are also shown in Fig. 2. On the addition of 6% woodash to the natural soil, the PI and LS remained almost the same while the FSI increased. However, the three properties started decreasing on the addition of 12% woodash attaining their minimum values on the addition of 18% woodash. Thus, PI, LS and FSI showed their highest improvement (decreased) on the addition of 18% woodash; PI decreased by 19.17%; LS deceased by 9.50% while FSI decreased by 2.15%.

These general improvements agree with Ismaiel (2006) and Okagbue (2007) in their use of fly ash and woodash for soil stabilization. According to Bell (1989, 1993), the improvement in the PI and LS is due to the instantaneous cation exchange occurring between the CaO contained in the woodash and the soil. As the woodash is added, its

Table 2. Grading, index properties and mineralogy of the expansive soil

Grading and index properties		Mineralogy	Abundance (%)
Sand (%)	49.00	Na-montmorillonite	7.85
Silt (%)	36.00	Illite	41.77
Clay (%)	15.00	Kaolinite	15.36
Liquid limit (%)	57.00	Sepiolite	23.65
Plastic limit (%)	26.84	Sanidine	11.35
Swell potential	8.80		
USCS*	CH		

* = Unified soil classification system

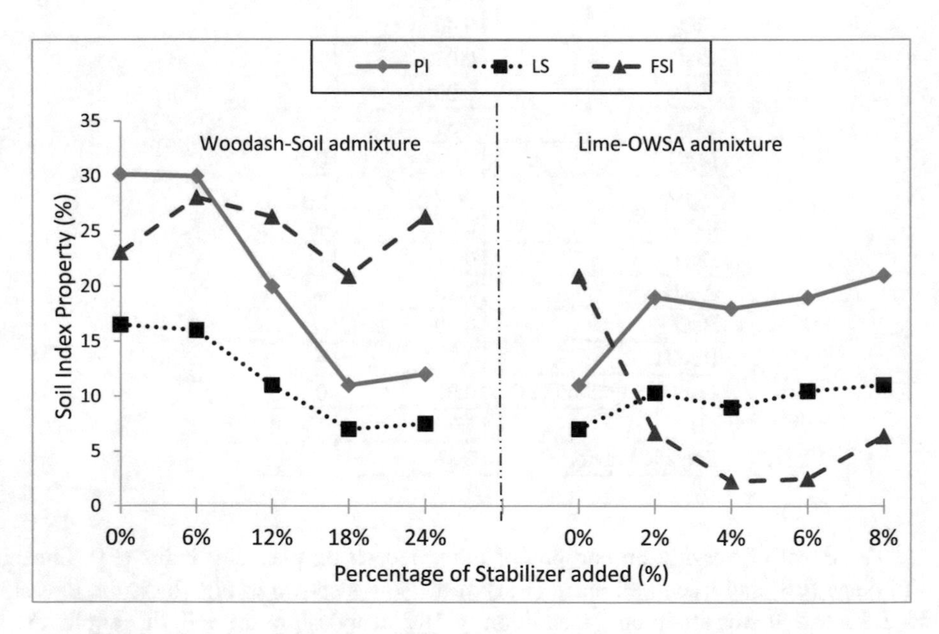

Fig. 2. Variation of PI, LS and FSI with varying proportions of the stabilizers

CaO content, in the presence of water, dissociates into Ca^{2+} and OH^{-} ions. The Ca^{2+} is absorbed by the soil particle surfaces - a process that results to flocculation and coagulation of the soil and consequent reduction in the plasticity (PI) and linear shrinkage (LS). In the present study, this process proves to be more effective on the addition of 18% woodash. However, due to the complexity in the chemistry of the used woodash (see Table 1), the addition of more woodash also increases other metallic ions (like K^{+}, Fe^{3+}, Mn^{2+}, Ag^{2+}), which inhibited the absorption of Ca^{2+} by the soil and thus increasing the PI and LS. The effect of the inhibition started on the addition of woodash in excess of 18% to the soil and continues even on the addition of lime to the OSWA as shown in Fig. 2. However, the inhibition seems ineffective to FSI and UCS of the soil on addition of lime to the OSWA.

FSI improved (decreased) by 2.15% on the addition of 18% woodash to the soil and further improved (decreased) by 18.66% on the addition of 4% lime to the OSWA. This is attributed to the dominant clay mineralogy of the soil and cation composition of the stabilizers. The higher valence Ca^{2+} contained in the woodash readily exchanges (replaces) the lower valence K^{+} and Na^{+} contained in the illite and Na-montmorillonite (see Table 2) thereby reducing the swellability of soil. This process continues through the addition of 18% woodash and also continues on the addition of lime to the OSWA up to 4%. It can be seen that the addition of woodash and lime in the excess of 18% and 4% respectively caused an increase in the FSI – a phenomenon that is consistent with results of earlier studies (Garg 2011).

Figure 3 shows that the addition of woodash to the soil did not cause significant improvement (increase) in the UCS. On the addition of 18% woodash, the soil UCS increased by 6.9 kPa (3.56% strength gain). Okogbue (2007) also used woodash in stabilizing soil but got a significant improvement of the UCS. This suggests that the chemical complexity of woodash used in present study also inhibited the supposed strength gain of the soil. The other cations (K^{+}, Fe^{3+}, Mn^{2+}, Ag^{2+}) contained in the woodash inhibited the strength improvement effect of the Ca^{2+}. However, the addition of 2% and 4% lime to the OSWA progressively increased the strength further by 387.40 kPa (193.12% strength gain). The lime suppressed the inhibition effect of other metallic cations. Therefore, the chemical complexity of the woodash inhibits the strength gain on the addition of woodash to the soil but has no effect to the strength on the addition of lime to the OSWA. Furthermore, Anifowose (1989) and Osula (1991) reported that strength gain is caused also by cation exchange reaction and is more effective between lime (CaO) and clay minerals than between lime and other minerals. The quantity of CaO contained in the stabilizer is commensurate with the strength gained by the soil. This explains why the strength gain is more when lime was added to the OSWA than when woodash was added to the natural soil.

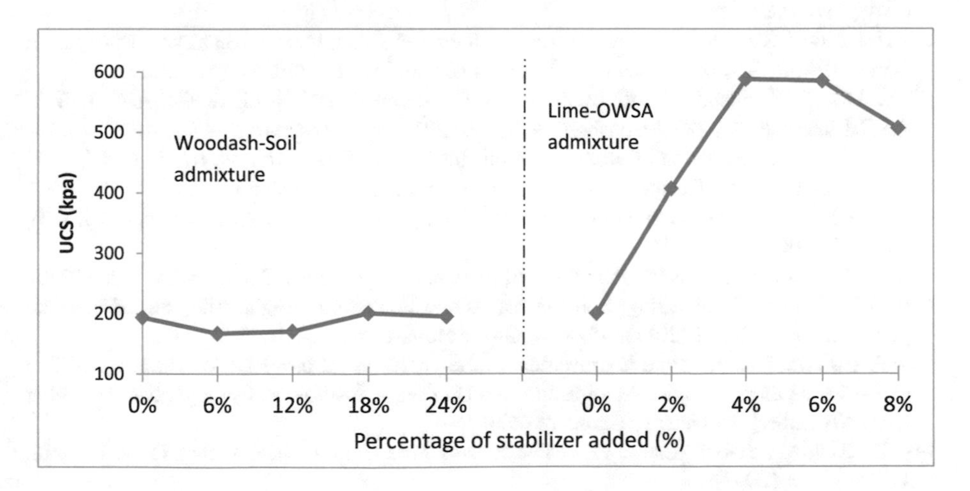

Fig. 3. Variation of UCS with varying proportions of the stabilizers

4 Conclusions

1. The stabilizing potential of woodash is, somewhat, dependent on its chemical composition. The chemical complexity of the woodash used in present study inhibited its stabilizing effect on the plasticity index (PI), linear shrinkage (LS) and unconfined compressive strength (UCS) of the soil in varying degrees but has no effect on the free swell index (FSI).
2. The addition of 18% woodash to 82% soil improved the PI and LS but has negligible effect on FSI and UCS. On the addition of 18% woodash to the soil, the PI decreased by 19.1%, LS decreased by 9.5% while FSI decreased by 2.1% and UCS increased by 3.5%.
3. The addition of lime to soil-woodash admixture did not improve the PI and LS but improved the FSI and UCS. Addition of 4% lime to the optimum soil-woodash admixture (OWSA) improved the FSI and UCS of the soil. The FSI further decreased by 18.66% and the UCS further increased by 193.12% (387.40 kPa).

References

Abubakar, M.B., Obaje, N.G.: Preliminary biostratigraphic evaluation of the hydrocarbon potentials of the cenomanian-turonian horizon in the gongila formation, upper benue trough Nigeria. J. Min. Geol. **37**(2), 121–128 (2001)

Agrawal, V., Gupta, M.: Expansive soil stabilization using marble dust. Int. J. Earth Sci. Eng. **4**(6), 59–62 (2011)

Anifowose, A.Y.B.: The performance of some soils under stabilization in Ondo state. Nigeria. Bull. Int. Assoc. Eng. Geol. **40**, 79–83 (1989)

ASTM C356-10: Standard Test Method for Linear Shrinkage of Preformed High-Temperature Thermal Insulation Subjected to Soaking Heat. ASTM International, West Conshohocken, PA (2010). www.astm.org

ASTM D2166/D2166 M-16: Standard Test Method for Unconfined Compressive Strength of Cohesive Soi., ASTM International, West Conshohocken, PA (2016). www.astm.org

ASTM D4546-14: Standard Test Methods for One-Dimensional Swell or Collapse of Soil. ASTM International, West Conshohocken, PA (2014). www.astm.org

Bell, F.G.: Lime stabilization of clay soils. Bull. Int. Assoc. Eng. Geol. **39**, 67–74 (1989)

Bell, F.G.: Engineering Treatment of Soils. Chapman and Hall, London (1993)

Brooks, R.M.: Soil stabilization with fly ash and rice husk ash. Int. J. Res. Rev. Appl. Sci. **1**(3), 209–217 (2009)

BS 1377-2: Methods of test for soils for civil engineering purposes. Classification tests (1990)

Garg, S.K.: Geotech Engineering: Soil Mechanics and Foundation Engineering, 8th ed. Khanna publishers, New Delhi (2011). 4575/15, Onkar House, Daryaganj

Ismaiel, H.A.H.: Treatment and improvement of the geotechnical properties of different soft fine grained soils using chemical stabilization. Ph.D. Thesis, Institute of Geology, Martin Luther Halle-Wittenberg University, Germany (2006)

Ji-ru, Z., Xing, C.: Stabilization of expansive soils by lime and fly ash. J. Wuhan Univ. Technol. **17**(4), 73–78 (2002)

Malhotra, M., Naval, S.: Stabilization of expansive soils using low cost materials. Int. J. Eng. Innovative Technol. **2**(11), 181–184 (2013)

Muntohar, A.S., Hantoro, G.: Influence of rice husk ash and lime on engineering properties of clayey subgrade. Electron. J. Geotech. Eng. (2000)

Murat, R.C.: Stratigraphy and Paleogeography of the Cretaceous and lower Tertiary in Southern Nigeria. African Geology. University of Ibadan Press, Ibadan (1972)

Ojoh, K.: Cretaceous geodynamics evolution of the southern part of the Benue Trough (Nigeria) in the equatorial domain of the south atlantic: stratigraphy, basin analysis and paleogeography. Bull Centers Resh. Explor. – Prod. EIF - Aquitaine **14**, 419–442 (1990)

Okagbue, C.O.: Stabilization of clay using woodash. J. Mater. Sci. Civ. Eng. **19**(1), 14–18 (2007)

Okagbue, C.O., Onyeobi, T.U.S.: Potential of marble dust to stabilise red tropical soils for road construction. Eng. Geol. **53**, 371–380 (1999)

Okogbue, C.O., Ugwoke, T.A.S.: Influence of petrogenesis on suitability of some pelitic rocks as construction aggregates in South-eastern Nigeria. Geot. Geol. Eng. **33**(6), 1395–1407 (2015)

Olade, M.A.: Evolution of Nigeria's benue trough (aulacogen): a tectonic model. Geol. Mag. **112**, 575–580 (1975)

Osula, D.O.A.: Lime modification of problem laterites. Eng. Geol. **30**, 141–153 (1991)

Rao, D.K., Rao, G.V.V.R., Pranav, P.R.T.: A laboratory study on the affect of rice husk ash and lime on the properties of marine clay. Int. J Eng. Innov. Tech. **2**(1), 345–353 (2012)

Show, K.Y., Tay, J.H., Goh, A.T.C.: Reuse of incinerator fly ash in soft soil stabilization. J. Mater. Civ. Eng. **15**(4), 335–343 (2003)

Experimental Investigation on the Compaction and Compressible Properties of Expansive Soil Reinforced with Bagasse Fibre and Lime

Liet Chi Dang(✉) and Hadi Khabbaz

School of Civil and Environmental Engineering, University of Technology Sydney (UTS), 15 Broadway, Ultimo, NSW 2007, Australia
{liet.dang,hadi.khabbaz}@uts.edu.au

Abstract. This paper presents a laboratory investigation into the mechanical characteristics of expansive soil reinforced with randomly distributed bagasse fibre and lime combination. Bagasse fibre, an agricultural waste by-product left after crushing of sugar-cane for juice extraction, was employed in this investigation as a reinforcing component for expansive soil reinforcement. Several series of laboratory experiments including standard compaction and consolidation tests were carried out on untreated soil and soil samples mixed with various contents of bagasse fibre in a wide range from 0% to 2% and a certain amount of 2.5% lime. The experimental results were used to comprehend the effects of adding bagasse fibre on the compaction and compressible properties of fibre reinforced soils with lime stabilisation. The compaction test results indicate that the addition of bagasse fibre, hydrated lime, and their combination decreased the dry density of stabilised soils. Moreover, the obtained results of the consolidation tests reveal that the reinforcement of expansive soil with bagasse fibre improved the pre-consolidation pressure, meanwhile tended to reduce the compression characteristics of the lime stabilised soils as bagasse fibre content increased from 0% to 1%. However, an excessive increase in bagasse fibre content beyond 1% to 2% was found to result in a slight reduction of the compressibility of lime-soil mixtures reinforced with bagasse fibre. The findings of this research provide a deeper insight into promoting applications of an agricultural waste by-product of bagasse fibre as a low-cost and eco-friendly material for treatment of expansive soils and fill materials for sustainable construction development in the field of civil infrastructure foundations.

1 Introduction

Expansive soils are fine-grained soil or decomposed rocks, showing significant volume change when exposed to variations of moisture content. Swelling and shrinkage behaviour is most likely to take place near the ground surface where it is directly prone to environmental and seasonal fluctuations. The expansive soils are usually at unsaturated state and have dominantly montmorillonite clay minerals. Most of the severe damage to residential buildings and other civil engineering structures built on top of expansive soils is dependent on the amount of monovalent cations absorbed into clay minerals. The average annual cost of damage to structures due to shrinkage and

J. S. McCartney and L. R. Hoyos (Eds.): GeoMEast 2018, SUCI, pp. 64–78, 2019.
https://doi.org/10.1007/978-3-030-01914-3_6

swelling is estimated about £400 million in the UK, $15 billion in the USA, and many billions of dollars worldwide (Jones and Jefferson 2012).

Stabilisation of expansive soil using chemical stabilisers (e.g. lime or cement) is commonly used as the most effective improvement method to overcome the adverse impacts of shrinkage and expansion behaviour of expansive soil because of its volume change again moisture variation. When lime is added into expansive soil, some physical and chemical changes between lime and expansive clay particles taking place in the presence of water alter the physico-chemical properties of expansive soil, which in turn changes the engineering characteristics of stabilised soil (Bell 1996; Dang et al. 2016c). Moreover, according to many researchers (Anggraini et al. 2016; Dang 2018; Dang et al. 2015a, b; Dang and Khabbaz 2018a, b, c; Dang et al. 2017b; Fatahi and Khabbaz 2013, 2015; Fatahi et al. 2012; Kampala and Horpibulsuk 2013), soil stabilisation using lime or cement combined with waste by-products (fly ash, bagasse ash, rice husk ash, coconut coir fibre, recycled carpet fibre and bagasse fibre, just to name a few) can extend the effectiveness of lime stabilisation of clayey soils in terms of compressive strength, shear strength, permeability, and ductility. Thus, the utilisation of combined lime and waste by-products was identified as an eco-friendly alternative solution in improving the engineering properties of clay expansive soils and it has increasingly become an extensive research interest in recent years. Moreover, it is interesting to state that as well-documented in the literature, lime stabilisation of clay expansive soil with fibre reinforcement can be used as an alternative earth load transfer platform in support of highway and railway embankments constructed on columns improved soft grounds (Dang et al. 2016a; 2017a; 2018a, b, c).

The experimental investigation by Chen and Indraratna (2014), however, indicated that conventional chemical stabilisers (e.g. lime, cement) for soil stabilisation are not always acceptable in Australia since they may cause adverse effects on the environment by changing the pH level of treated soil and its surrounding areas. As a result, the quality of ground-water and the normal growth of vegetation can be affected because of the pH change. On top on that, the increasing use of conventional chemical agents to stabilise soil can produce high compressive strength, but also increase the brittleness behaviour of stabilised soil, which influences the soil stability when subjected to cyclic traffic loading under road and railway embankments. Therefore, an environmentally friendly alternative solution such as bagasse fibre reinforcement of soils combined with lime stabilisation is necessary to improve the strength, the ductility and the durability of stabilised soils, meanwhile minimises the negative effects on the environment.

Recently, Dang et al. (2016b) examined the performance of bagasse fibre in enhancing the linear shrinkage and the compressive strength of compacted expansive soils by changing bagasse fibre content from 0% to 2% along with increasing curing time from 3 to 28 days. The test results indicated that as the curing time increased up to 7 days, the introduction of bagasse fibre reinforcement from 0% to 2% improved both the linear shrinkage and the compressive strength of expansive soil and then they remained most likely unchanged with a longer curing time. They concluded that the improvement in the shrinkage and the strength of soils reinforced with bagasse fibre might be due to the development of interaction and interlocking mechanism between fibre surface and soil matrix by compaction energy and with time. Viswanadham et al. (2009) studied the swelling behaviour of geofibre reinforced expansive soil by mixing various polypropylene fibre content from 0.25% to 0.5% and different aspect ratios of 15, 30 and 45. Based on the favourable results obtained, they concluded that polypropylene fibre effectively improved the heave and swelling pressure of expansive soil. The maximum reduction of the heave and the swelling pressure was found at the fibre content of 0.25% and the lower aspect ratio between 15 and 30. According to Mohamed (2013), the shear strength and tensile strength of expansive clay soil reinforced with Hay fibre increased as the fibre content added into soil mixtures increased up to 1%. Meanwhile, the shrinkage limit and the swelling potential decreased up to 1% Hay fibre insertion to the soil matrix followed by an increase with higher fibre addition up to 1.5%. Although those aforementioned experimental investigations indicated that both natural and synthetic fibre could be beneficial for the engineering property improvement of expansive soil with or without chemical stabilisation as fibre content increased from 0% to 1.5%. However, the influence of a combination of natural fibres such as bagasse fibre and lime for expansive soil treatment on the other engineering properties such as compaction and compressibility have not fully been investigated and well reported in the literature.

To have a comprehensive understanding of the potential utilisation of natural fibre for expansive soil reinforcement, several series of laboratory experiments, including standard compaction and consolidation tests, were performed on natural expansive soil and treated soil samples with different contents of randomly distributed bagasse fibre ranging from 0 and 2% and a fixed lime content after 7 days of curing and soaking. The test results of this experimental investigation are analysed and discussed to comprehend the influence of bagasse fibre reinforcement and lime stabilisation on the compaction and compression characteristics of expansive soil. It should be noted that the only results obtained from the compaction and consolidation tests are presented in this paper, which are part of an ongoing research project of characterisation and treatment of expansive soils using agricultural waste by-products (bagasse ash and fibre). Further experimental evaluations of the influence of bagasse fibre inclusions on the shrink-swell behaviour of reinforced expansive soils could be found in Dang et al. (2016b; 2017c). Moreover, as known that natural fibre reinforcement is biodegradable with time, it is indispensable to improve the durability of natural fibre such as bagasse fibre by applying chemical treatment (i.e. sodium hydroxide, sodium silica, sodium sulphite) and/or coating (asphalt emulsion, rosin-alcohol, acrylic, polystyrene, and silane) to prevent water absorption. However, the results of those tests, which are beyond the scope of this paper, were identified as a follow-up publication.

2 Materials

2.1 Natural Soil

Soil samples collected from Queensland, Australia, was used in this experimental investigation. After removal of visible organic matters such as tree roots and leaves, the soil was air-dried and broken into pieces in the laboratory. The specific gravity of soil solids (G_s) was 2.64 ± 0.02. The grain size distribution illustrated that there were 0.1% of particles in the range of gravel, 18.3% in the range of sand and 81.6% of fine-grained material (i.e. silt/clay). Atterberg limits of the fine-grained portion of material were about 86% liquid limit (LL) and 37% plastic limit (PL), which yielded to a plasticity index (PI) of 49%. The average linear shrinkage and natural moisture content of the samples was 21.7% and 30.8%, respectively. In term of sizes of particles, the soil was classified as high plasticity clay (CH) according to the Unified Soil Classification System (USCS) (AS 1993). Based on the high linear shrinkage and plasticity index, the soil can be classified as highly expansive soil.

2.2 Bagasse Fibre

Bagasse fibre used in this study was obtained from ISIS Central Sugar Mill Co., Ltd, Queensland in Australia. The bagasse fibre, as depicted in Fig. 1, had a diameter ranging from 0.3 mm to 3.1 mm and a length ranging from 0.3 mm to 13.8 mm. The specific gravity of bagasse fibre (G_f) was about 1.25–1.55 and their average tensile strength was 96.24 ± 29.95 MPa. The obtained fibre was air dried in a controlled room environment with a temperature of 25 °C and a relative humidity of 80% until its mass remained constant. Then, the dried fibres were carefully sieved and passed through 9.5 mm aperture sieve and retained on 300 μm aperture sieve, which were selected for this investigation.

Fig. 1. Bagasse fibre used in this investigation

2.3 Hydrated Lime

Hydrated lime utilised in this study (Fig. 2) has about 90% of calcium hydroxide. The hydrated lime was locally purchased in Sydney. Table 1 shows the physical and chemical properties of hydrated lime provided by the producer.

Fig. 2. Hydrated lime used in this investigation

Table 1. Chemical composition and physical properties of hydrated lime

Physical properties		Chemical composition	
Property	Value	Components	Content (%)
Specific gravity	2.2-2.3	Ignition loss	24%
Bulk density (kg/m^3)	400-600	SiO_2	1.8
pH	12.0	Al_2O_3	0.5
		Fe_2O_3	0.6
		CaO	72.0
		MgO	1.0
		CO_2	2.5

3 Sample Preparation and Experimental Program

3.1 Mixing of Materials

Soil samples were prepared by thoroughly mixing the pulverized natural soil with individual hydrated lime, bagasse fibre or their combination, as shown in Table 2, to become homogeneous mixtures before tap water was added at the target water content. It can be noted that in this investigation, the additive and water contents were calculated based on the total dry weight of each mixture. Following this preparation, the mixtures were mixed thoroughly using a mechanical mixer. After mixing of the materials, soil specimens (Fig. 3) were prepared for many conventional geotechnical experiments.

3.2 Standard Compaction Test

Standard compaction test was carried out to determine the maximum dry density (MDD) and the optimum water content (OMC) for untreated and treated soils in accordance with the procedures prescribed in AS 1289.5.1.1 (AS 2017). Different water

Table 2. Summary of mixtures employed in this investigation

Mix no.	Bagasse fibre content (%)	Hydrated lime (%)	Notes
1	0	0	Natural soil
2	0.5	0	Bagasse fibre and soil
3	1.0	0	
4	2.0	0	
5	0.5	2.5	Lime, bagasse fibre and soil
6	1.0	2.5	
7	2.0	2.5	

Fig. 3. Additives and soil mixing

contents were added to the pulverised soils and thoroughly mixed to make them uniformly distributed through the soil mixtures. Then, the soil mixtures were placed in a cylindrical metal mould, with an internal diameter of 105 mm and height of 115.5 mm and compacted in three equal layers by 25 uniformly distributed blows on the rammer falling freely from a height of 300 mm in accordance with the procedure of AS 1289.5.1.1 (AS 2017). The specimens (Fig. 4) were extruded, measured and weighed, and their moisture contents were determined. The dry density and water content of untreated soil for each specimen were calculated and recorded in accordance with AS 1289.5.1.1 (AS 2017). A total of eleven tests on samples with different water contents were conducted to determine the MDD and the OMC of untreated soil. After that, different amounts of additives, as shown in Table 2, were added to the pulverised soil at the optimum moisture content of untreated soil. The blended mixtures were compacted in the same procedure applied to the untreated soil. The dry density of each mixture was achieved and used for carrying out other geotechnical engineering tests.

Fig. 4. Compacted soil sample

3.3 One-Dimensional (1D) Consolidation Tests

A series of one-dimensional swelling consolidation tests were carried out on untreated and treated soil specimens after 3 days of curing using conventional Oedometer apparatus following the testing procedure in accordance with AS 1289.6.6.1 (AS 1998). For sample preparation, soil specimens were compacted in a cylindrical metal ring, with an internal diameter of 50 mm and height of 17 mm, at the MDD and OMC. After specimen preparation and curing for 3 days, the compacted soil was extruded and put into the Oedometer ring of the same diameter and then the standard Oedometer testing setup was followed (two-way drained setup). An initial seating load of 3 kPa was applied. Once proper loading contact was achieved, the sample was inundated with distilled water and remained for 4 days to get full saturation prior to compression. During the saturation stage of the consolidation tests, vertical swell deformations were measured using a dial gauge. After completion of this stage, the soil specimens were subjected to additional pressure incrementally in accordance with the standard consolidation test procedure AS 1289.6.6.1 (AS 1998). At any level of pressure, the applied pressure was kept on the specimens for 24 h to ensure the completion of consolidation. For each type of mixtures, at least two samples were tested and then the average of the compression test results was presented.

4 Results and Discussion

4.1 Influence of Additive Content on Compaction Characteristics of Expansive Soil

The standard compaction curve of untreated expansive soil as a preliminary step to determine the maximum dry density (MDD) and the optimum moisture content (OMC) of the soil only is presented in Fig. 5a. Analysis of the compaction results of natural soil found the MDD and the OMC to be 12.9 kN/m^3 and 36.5%, respectively. Subsequently, several series of soil samples mixed soils with different contents of bagasse fibre and lime were prepared at the OMC of the untreated soil (36.5%) to investigate the influence of the additive content on the so-called maximum dry density of the treated soil mixtures. The obtained results of the standard compaction tests for expansive soil reinforced with various contents of bagasse fibre reinforcement and lime are depicted in Fig. 5b. It is observed that with inclusion of bagasse fibre into soils without lime treatment, the MDD of the reinforced soil mixtures gradually decreased from 12.9 kN/m^3 to 11.4 kN/m^3 as bagasse fibre content increased from 0% to 2%. The MDD reduction could be due to the lower specific gravity of bagasse fibre in comparison with that of untreated soil. Furthermore, addition of 2.5% hydrated lime into bagasse fibre reinforced soil mixtures as observed in Fig. 5b shows that the MDD of stabilised soils decreased further. It is noted that the MDD of each lime-bagasse fibre-soil mixture was obviously lower than that of the bagasse fibre reinforced soils without lime treatment. The MDD decrease of lime treated soils with bagasse fibre might be attributed to the lower specific gravity of bagasse fibre together with the flocculation and agglomeration because of cation exchange processes between clay particles and lime that changed the soil particles to be coarser particles. As mentioned earlier, the formation of the coarser particles occupying the larger spaces in the soil matrix, increased the void volume and hence reduced the dry density of the treated soil mixtures. In addition, tiny air gaps trapped into fibre surface could be another possible reason that explains the reduction of the MDD of lime-fibre-mixtures. An increase in bagasse fibre content led to the increase in the tiny air gaps and hence reduced the MDD of stabilised soil composite. Results of this investigation are consistent with other researcher observations (Ayeldeen and Kitazume 2017; Kinuthia et al. 1999).

4.2 Effects on the Compression Characteristics

The previous investigation by the authors revealed that for only bagasse fibre reinforcement of expansive soil, when bagasse fibre content increased from 0% to 2%, the unconfined compressive strength, the California bearing capacity, and the linear shrinkage of reinforced soils, compared to non-reinforced soil, were found to significantly improve by approximately 40%, 34% and 40%, respectively. However, the swelling potential was observed to reduce by a great amount of 48% as bagasse fibre content increased from 0% to 1% and this was followed a small increase in the swelling potential as an additional increase in bagasse fibre exceeded 1%. These findings confirmed the effectiveness of adding bagasse fibre alone in improving the shrink-swell behaviour and the other mechanical characteristics of reinforced expansive soils. Further details of the experimental investigations could be found in Dang et al. (2016b; 2017c).

In this study, a series of Oedometer tests was undertaken on expansive soils treated with combinations of a fixed lime content with different contents of bagasse fibre after 3 days of curing and 4 days soaking (so-called 7 days of curing). As it can be found in Fig. 6 that when the bagasse fibre content was varied from 0% to 2%, only 2.5% hydrated lime were added to bagasse fibre-soil mixtures to study the effect of adding bagasse fibre on the compressible property of lime treated expansive soil. Moreover, to

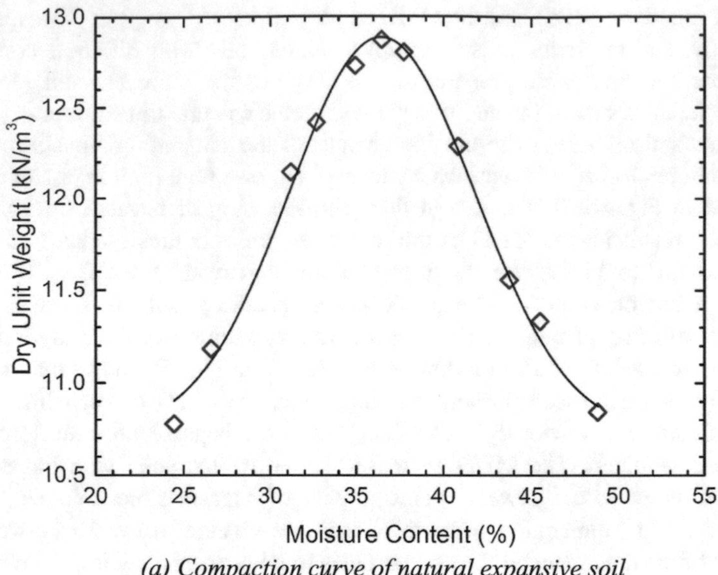

(a) Compaction curve of natural expansive soil

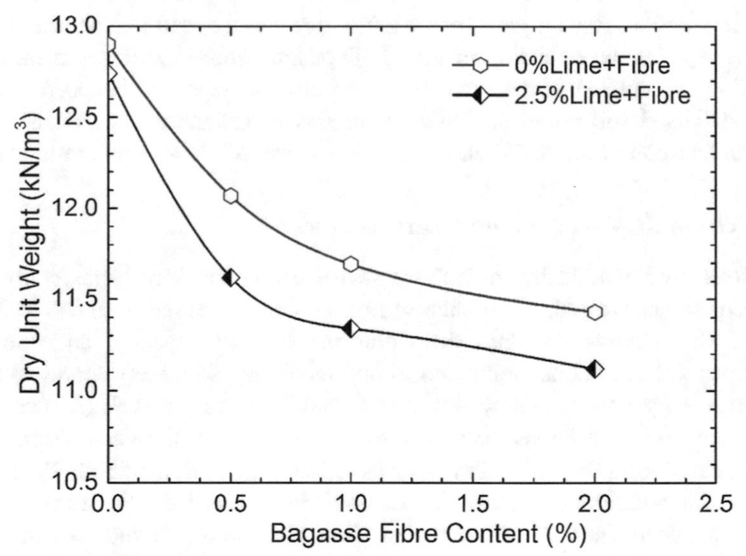

(b) Compaction curves of lime treated expansive soil with bagasse fibre reinforcement

Fig. 5. Compaction curves of expansive soil reinforced with different contents of bagasse fibre and hydrated lime

have a better comparison of the compressibility between untreated soil and lime treated soils with bagasse fibre reinforcement, the test results obtained from untreated soil sample is also depicted in Fig. 6. Observation of the experimental results in Fig. 6

reveals that adding bagasse fibre into lime treated soil mixtures was found to reduce the compression characteristics of reinforced soils as bagasse fibre content increased from 0% to 1%. Subsequently, the compressibility of lime treated soils reinforced with bagasse fibre indicated a slight increase when the addition of bagasse fibre exceeded 1%. When compared with the compression curves of untreated soil and 2.5% lime treated soil, the lower slope reduction of the virgin compression curves was found for soil samples treated with 2.5% lime in combination of bagasse fibre reinforcement. Referring to Fig. 6, the combination 2.5% lime and 1% bagasse fibre caused the most notable improvement in the virgin curve of reinforced soils.

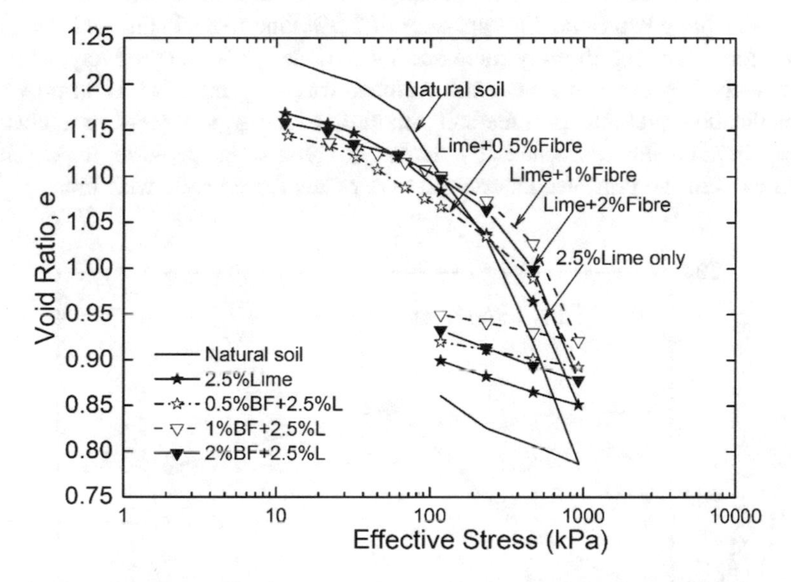

Fig. 6. Compression curves of expansive soil reinforced with different contents of bagasse fibre and 2.5% lime

The compressibility decrease of expansive soil stabilised with lime-bagasse fibre combination could be attributed to replacement of exchangeable ions on the clay surface with calcium ions in lime as a result of cation exchange. This phenomenon transformed the clay particles in the lime-soil matrix to become coarser, stronger and less plastic, which consequently promoted the improvement in the compressibility of treated soils. In addition, when bagasse fibre was introduced into the lime-soil mixture, interactions between bagasse fibre surface and lime-soil matrix with curing time might contribute to additional improvement in the compressible property of stabilised soils. As expected, when bagasse fibre content increased, the fibre surface area that was exposed to soil matrix also increased, which facilitated the better resistance to the compression pressure. However, the excessive addition of fibre bagasse content into lime-soil mixtures was about to increase the compressibility of stabilised soils to a certain extent due to the relatively high compressibility of natural bagasse fibre compared with soil particles.

The influence of bagasse fibre reinforcement on the preconsolidation pressure of lime treated expansive soil is illustrated in Fig. 7. It is noted that the preconsolidation pressures of 2.5% lime treated soils with various bagasse fibre contents were deprived from Fig. 6 using a proposed method of Boone (2010). Observation of the preconsolidation pressures presented in Fig. 7 notes that the increase in bagasse fibre content from 0% to 1% to reinforce soils with 2.5% lime resulted in the corresponding increase in the pre-consolidation pressure from 165 kPa to around 240 kPa (approximately 45% improvement). When additional increase in bagasse fibre content beyond 1% exhibited a marginal decrease in the preconsolidation pressure of reinforced soils to a certain value of 215 kPa. However, by comparing with untreated soil and only 2.5% lime treated soils, the preconsolidation pressure of 2.5% lime treated soil with 2% bagasse fibre reinforcement significantly increased by 38% and 30%, respectively. This behaviour reveals that bagasse fibre reinforcement was very effective in improving the preconsolidation pressure of lime-soil mixture, whereas an excessive inclusion of bagasse fibre content exceeded 1% tented to reduce its positive impact on the improvement in the preconsolidation pressure of reinforced soils with lime.

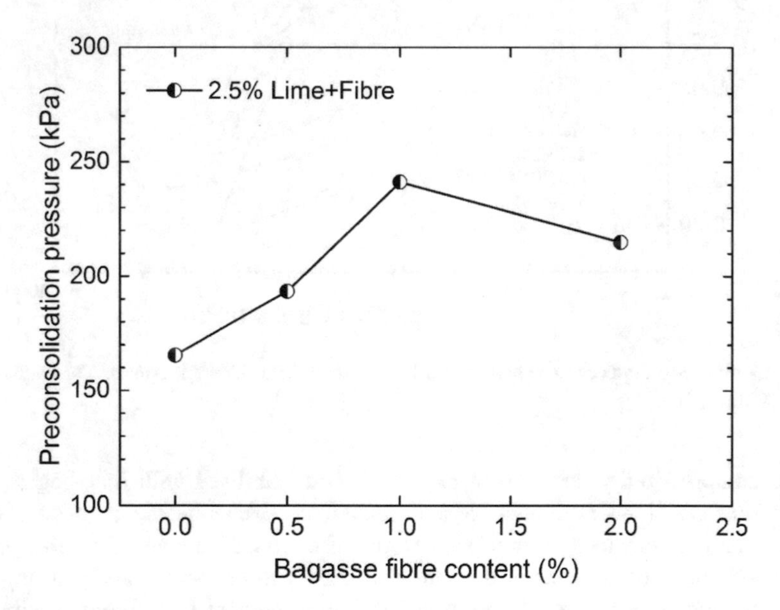

Fig. 7. Variation of preconsolidation pressure of expansive soil reinforced with various contents of bagasse fibre and 2.5% lime

Figure 8 indicates the influence of bagasse fibre reinforcement on the variation of compression index (Cc) and swelling index (Cs) of 2.5% lime-soil mixtures after curing and soaking for 7 days. It can be noted that Cc is defined as the slope of the straight line portion (virgin compression portion) of the effective stress-void ratio curve, meanwhile

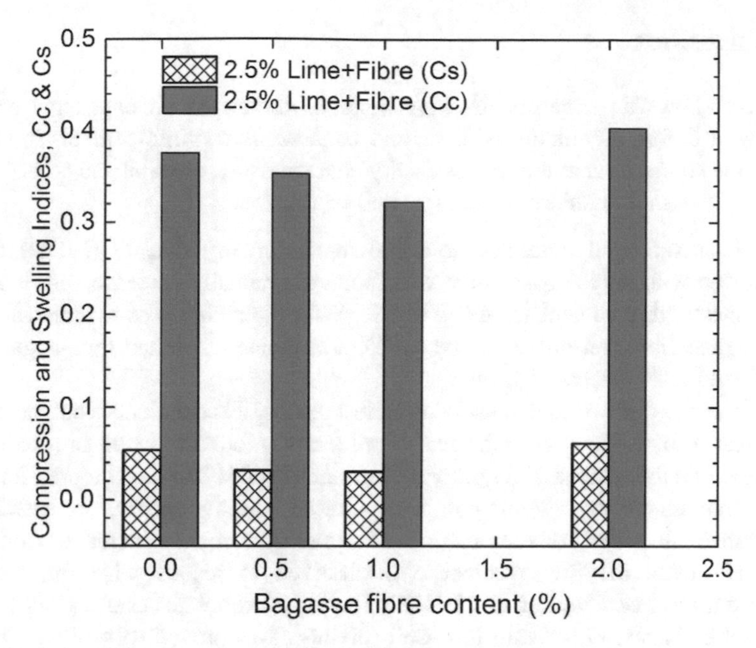

Fig. 8. Variations of compression and swelling indices of expansive soil reinforced with various contents of bagasse fibre and 2.5% lime

Cs is the slope of the unloading compression curve. As can be seen in Fig. 8, the value of Cc appeared to reduce with increasing bagasse fibre inclusion from 0% to 1%. However, an increasing trend of the compression index was observed when the bagasse fibre inclusion into the lime-soil mixture increased beyond 1%. A similar behaviour can be found the swelling index of lime treated soils with bagasse fibre reinforcement as bagasse fibre content increased up to 2%. As observed in Fig. 8, the change of Cc was more pronounced than the Cs variation of reinforced soils with lime stabilisation. The reduction of both the Cc and Cs indices confirms that the addition of bagasse fibre can effectively reduce the compressibility of lime treated soils. As noted earlier, the improvement of lime treated soils reinforced with bagasse fibre might be due to the interlocking mechanism and interaction between lime-soil matrix and bagasse fibre surface that play an important role in improving the mechanical properties of reinforced soils. As bagasse fibre content increased to a certain amount, the lime-soil mixtures with bagasse fibre reinforcement would promote the better resistance to the applied compression pressure, and consequently facilitate the lower compressibility of reinforced soils.

5 Conclusions

This paper shows an experimental investigation, conducted on expansive soils stabilised with different contents of lime and bagasse fibre reinforcement, in order to evaluate the compaction and compressibility characteristics of stabilised soils. The key findings of this investigation are summarised as follows:

- In comparison with untreated soil, the maximum dry density (MDD) of soils reinforced with only bagasse fibre was found to gradually decrease with increasing the bagasse fibre content from 0% to 2%. With the addition of hydrated lime into the bagasse fibre-soil mixtures, the MDD of combined hydrated lime-bagasse fibre reinforced soils decreased further.
- From the one-dimensional consolidation test results, it is concluded that the addition of bagasse fibre into soils stabilised with lime was found to result in a remarkable influence on the compressible properties of reinforced soils. The compression curve, the compression and swelling indices reduced, meanwhile the preconsolidation pressure of stabilised soils improved, as bagasse fibre content increased from 0% to 1%. However, the improvement was observed to reduce when bagasse fibre inclusion increased further to 2%. This finding corroborates that adding a certain amount of bagasse fibre into lime-soil mixtures was proved to promote the most effective improvement in the compressibility characteristics of treated soil.
- This investigation indicated that the utilisation of hydrated lime-bagasse fibre combination for expansive soil treatment could highly be effective in not only improving the geotechnical properties of expansive soil but also minimising the environmental impacts of an agricultural waste by-product of bagasse fibre. This study also revealed that bagasse fibre reinforcement had the potential use as a recycled, environmentally friendly and cost-effective additive in combination of with lime for sustainable civil infrastructure construction development because of reducing the consumption of conventional stabilisers such as lime or cement, commonly adopted in treatment of expansive soil.

Acknowledgments. The results presented in this paper are part of an ongoing research at University of Technology Sydney (UTS) supported by the Australian Technology Network (ATN), Arup Pty Ltd., Queensland Department of Transport and Main Roads (TMR), ARRB Group Ltd and Australian Sugar Milling Council (ASMC). The authors gratefully acknowledge their support.

References

Anggraini, V., Asadi, A., Farzadnia, N., Jahangirian, H., Huat, B.: Reinforcement benefits of nanomodified coir fiber in lime-treated marine clay. J. Mater. Civil Eng. 06016005 (2016)

AS 1993: Geotechnical Site Investigations, AS 1726-1993, Standards Australia, Sydney, Australia

AS 1998: Determination of the one-dimensional consolidation properties of a soil - Standard method, AS 1289.6.6.1-1998, Standards Australia, Sydney, Australia

AS 2017: Determination of the dry density/moisture content relation of a soil using standard compactive effort, AS 1289.5.1.1:2017. Standards Australia, Sydney, Australia
Ayeldeen, M., Kitazume, M.: Using fiber and liquid polymer to improve the behaviour of cement-stabilized soft clay. Geotext. Geomembr. **45**(6), 592–602 (2017)
Bell, F.G.: Lime stabilization of clay minerals and soils. Eng. Geol. **42**(4), 223–237 (1996)
Boone, S.J.: A critical reappraisal of "preconsolidation pressure" interpretations using the oedometer test. Can. Geotech. J. **47**(3), 281–296 (2010)
Chen, Q., Indraratna, B.: Shear behaviour of sandy silt treated with lignosulfonate. Can. Geotech. J. **52**(8), 1180–1185 (2014)
Dang, L.C.: Enhancing the engineering properties of expansive soil using bagasse ash. Bagasse Fibre and Hydrated Lime, Ph.D. thesis, University of Technology Sydney (2018)
Dang, L.C., Dang, C., Fatahi, B., Khabbaz, H.: Numerical assessment of fibre inclusion in a load transfer platform for pile-supported embankments over soft soil. In: Chen, D., Lee, J., Steyn, W.J. (eds.) Geo-China 2016, vol. GSP 266, pp. 148–155. ASCE (2016a)
Dang, L.C., Dang, C.C., Khabbaz, H.: Behaviour of columns and fibre reinforced load transfer platform supported embankments built on soft soil. In: The 15th International Conference of the International Association for Computer Methods and Advances in Geomechanics, Wuhan, China (2017a)
Dang, L.C., Dang, C.C., Khabbaz, H.: Modelling of columns and fibre reinforced load transfer platform supported embankments. Ground Improvement (Under Review) (2018a)
Dang, L.C., Dang, C.C., Khabbaz, H.: Numerical analysis on the performance of fibre reinforced load transfer platform and deep mixing columns supported embankments. In: Bouassida, M., Meguid, M.A. (eds.) Ground Improvement and Earth Structures, pp. 157–169. Springer, Cham (2018b)
Dang, L.C., Dang, C.C., Khabbaz, H.: A parametric study of deep mixing columns and fibre reinforced load transfer platform supported embankments. In: Advances in Foundation and Ground Improvement Techniques. Springer, Cham (2018c)
Dang, L.C., Fatahi, B., Khabbaz, H.: Behaviour of expansive soils stabilized with hydrated lime and bagasse fibre. Procedia Eng. **143**, 658–665 (2016b)
Dang, L.C., Hasan, H., Fatahi, B., Jones, R., Khabbaz, H.: Effects of bagasse ash and hydrated lime addition on engineering properties of expansive soil. In: GEOMATE 2015, Osaka, Japan, pp. 90–95 (2015a)
Dang, L.C., Hasan, H., Fatahi, B., Jones, R., Khabbaz, H.: Enhancing the engineering properties of expansive soil using bagasse ash and hydrated lime. Int. J. GEOMATE **11**(25), 2447–2454 (2016c)
Dang, L.C., Hasan, H., Fatahi, B., Khabbaz, H.: Influence of bagasse ash and hydrated lime on strength and mechanical behaviour of stabilised expansive soil. In: GEOQuébec 2015, Québec City, Canada (2015b)
Dang, L.C., Khabbaz, H.: Assessment of the geotechnical and microstructural characteristics of lime stabilised expansive soil with bagasse ash. In: GeoEdmonton 2018, Alberta, Canada (2018a)
Dang, L.C., Khabbaz, H.: Enhancing the strength characteristics of expansive soil using bagasse fibre. Springer Series in Geomechanics and Geoengineering. Springer, Cham (2018b)
Dang, L.C., Khabbaz, H.: Shear strength behaviour of bagasse fibre reinforced expansive soil. In: IACGE2018, Geotechnical Special Publications. ASCE, Chongqing, China (2018c)
Dang, L.C., Khabbaz, H., Fatahi, B.: Evaluation of swelling behaviour and soil water characteristic curve of bagasse fibre and lime stabilised expansive soil. In: PanAm-UNSAT 2017, ASCE, Texas, USA (2017b)

Dang, L.C., Khabbaz, H., Fatahi, B.: An experimental study on engineering behaviour of lime and bagasse fibre reinforced expansive soils. In: 19th ICSMGE, Seoul, Republic of Korea, pp. 2497–2500 (2017c)

Fatahi, B., Khabbaz, H.: Influence of fly ash and quicklime addition on behaviour of municipal solid wastes. J. Soils Sediments **13**(7), 1201–1212 (2013)

Fatahi, B., Khabbaz, H.: Influence of chemical stabilisation on permeability of municipal solid wastes. Geotech. Geol. Eng. **33**(3), 455–466 (2015)

Fatahi, B., Khabbaz, H., Fatahi, B.: Mechanical characteristics of soft clay treated with fibre and cement. Geosynthetics Int. **19**, 252–262 (2012)

Jones, L.D., Jefferson, I.: Expansive soils. ICE Manual of Geotechnical Engineering, pp. 413–441. ICE Publishing, London (2012)

Kampala, A., Horpibulsuk, S.: Engineering properties of silty clay stabilized with calcium carbide residue. J. Mater. Civil Eng. **25**(5), 632–644 (2013)

Kinuthia, J.M., Wild, S., Jones, G.I.: Effects of monovalent and divalent metal sulphates on consistency and compaction of lime-stabilised kaolinite. Appl. Clay Sci. **14**(1), 27–45 (1999)

Mohamed, A.E.M.K.: Improvement of swelling clay properties using hay fibers. Constr. Build. Mater. **38**, 242–247 (2013)

Viswanadham, B.V.S., Phanikumar, B.R., Mukherjeeb, R.V.: Swelling behaviour of a geofiber-reinforced expansive soil. Geotext. Geomembr. (2009)

Mechanical and Hydraulic Properties of Bentonite Clay Stabilized with Cement, Lime, and Mixed Lime-Cement by Dry and Wet Methods After 5 Years of Curing

M. Farzi(✉) and Mohammad S. Pakbaz

Department of Civil Engineering,
Shahid Chamran University of Ahvaz, Ahvaz, Iran
Mohsen.farzi@gmail.com, Pakbaz_m@scu.ac.ir

Abstract. Stabilization and improvement of soil quality in terms of increased compressive strength and reduced settlement and other geotechnical parameters have been long investigated. Various studies have examined the effects of different types and amounts of stabilizers on different soils. In most of these studies, fine-grained soils have been tested at optimal humidity to study the effect of the stabilizers, with the impact of stabilizers at liquid limit been rarely considered. One of the most important issues in this regard is the long-term effect of stabilizers on the trend of variations in geotechnical parameters. However, to the best of our knowledge, no study has been done on the long-term effect of mixing method (wet and dry) on bentonite clays. Accordingly, in the present research, bentonite clay was stabilized with different percentages of cement and lime at liquid limit of the samples by dry and wet methods. The samples were retested after about 5 years to measure the long-term effect of the mixing method.

1 Introduction

Quality improvement of geotechnical soil properties has been long investigated by many researchers and engineers in this field. Lime and cement are among the oldest soil stabilizers, such that they are known as traditional stabilizers. Many researchers have examined the effects of these stabilizers on mechanical and hydraulic properties of fine-grained soils. According to the results of these studies, being based on pozzolanic reactions, soil cementation may last months or even years (Wild et al. 1998). One of the parameters positively contributing into compressive strength of soil samples is the type of minerals composing the soil (Vitale et al. 2017). The effects of curing time and temperature have been also addressed in some studies whose results revealed relationships between these two parameters, in one hand, and compressive strength of the samples stabilized by lime and cement, on the other hand (Bell 1996). Investigation of curing time and compressive strength of the lime- and cement-stabilized samples in several years indicates the relationship between the increasing trends of compressive strength of samples and different contents of the stabilizers, i.e. the higher the percentage of stabilizer, the higher the compressive strength, even after 5 years (Nagaraj

J. S. McCartney and L. R. Hoyos (Eds.): GeoMEast 2018, SUCI, pp. 79–84, 2019.
https://doi.org/10.1007/978-3-030-01914-3_7

et al. 2014). It must be noted that, the presented results in this paper somehow complete the results of Pakbaz and Farzi (2015) after about five years of curing.

2 Materials and Methods

For further investigation of the effects of stabilizers such as cement, lime, and mixed lime-cement on mechanical and hydraulic properties of bentonite clay prepared by two methods (wet and dry), 7, 14 and 28-year samples were studied. The samples were originally investigated by Pakbaz and Farzi (2015). Obtained results indicated the effect of mixing method on mechanical and hydraulic properties of the bentonite clay. As the effects of the additives incorporated into the two mixing methods (dry and wet) on soil samples are yet to be addressed, in this study, the samples stabilized with cement and lime (by wet and dry preparation methods) were evaluated after 1850 days of curing using unconfined compressive strength and consolidation tests.

3 Results and Discussion

3.1 Unconfined Compressive Strength Test

Subjecting the 1850-day samples to unconfined compressive strength test, the results of all samples are listed on Table 1. In a previous paper, the authors gave information about the samples 7, 14 and 28.

Regarding the obtained results, increased compressive strength was observed in all the samples, with the highest strength experienced in the samples containing 10% of cement (as stabilizer) prepared by wet method. Compared to 28-day samples, increased percentages of compressive strength for all 1850-day samples are tabulated in Table 2.

Regarding Table 2, the maximum increase in compressive strength was seen in the samples containing cement as stabilizer. A comparison between the wet and dry samples indicated that the samples prepared by dry method had the highest growth in compressive strength.

3.2 One-Dimensional Consolidation Test

All of 1850-day samples were subjected to one-dimensional consolidation test. Regarding the curing time and increased compressive strength of the samples, the samples containing 6, 8, and 10% of additives showed high resistance against settlement, as indicated by very low settlement values. Therefore, results of the consolidation test were investigated only for the samples containing 2 and 4% of additives. Overall results of all samples against the applied loads were fully similar to those of the 28-day samples.

Results of the consolidation tests on the samples stabilized with cement by the two methods (dry and wet) are depicted in Fig. 1.

Table 1. All the test results unconfined compressive stress (kPa)

Unconfined compressive stress (kPa)	Additive	Curing day	Dry method					Wet method				
			Additive percent									
			2	4	6	8	10	2	4	6	8	10
	Cement	7	10/53	22/24	67/65	139/44	197/07	12/21	55/94	158/55	311/23	411/49
		14	11/06	26/33	113/85	170	237/85	14/34	58/59	173/42	315/92	422/41
		28	18/03	63/58	128/95	335/87	449/53	19/72	64/43	184/38	358/99	495/15
		1850	19	74	144	372	490	22	79	200	385	540
	Lime	7	19/24	51/41	145/51	320/97	417/56	20/37	55/55	187/8	260/57	396/26
		14	30/48	70	152/69	325/01	421/51	21/34	56/28	198/56	275/16	403/09
		28	37/24	119/37	162/96	345/33	453/84	22/31	58/24	202/13	326/11	415/5
		1850	46/3	131	177	370	480	24	66/6	220	342	435
	Cement-Lime	7	8/61	52/83	202/68	354/08	440/69	6/78	43/09	193/1	329/92	435/72
		14	16/45	67/39	209/22	382/76	443/53	14/66	55/64	206/34	349/76	437/98
		28	21	68/16	217/6	407/08	455/17	17/47	61/6	212/55	362/89	452/69
		1850	25	73	232	435	481	22	86	221	386	470

Table 2. Percent increase in unconfined compressive stress after 1850 day in comparison with 28 day samples

Additive	Dry method					Wet method				
	Additive percent									
	2	4	6	8	10	2	4	6	8	10
Cement	5.4	16.4	11.7	9	11.6	22.6	7.2	8.47	7.2	9
Lime	24.3	9.7	8.6	7	5.76	7.6	14.4	8.8	4.9	4.7
Cement-Lime	19	7	6.6	6.9	5.7	8.6	13.6	4	6.4	3.8

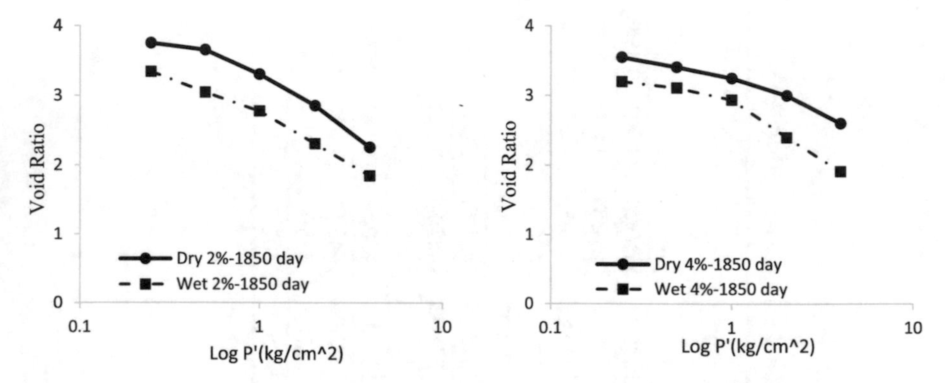

Fig. 1. End of primary e-log p′ relationship for cement treated samples after 1850 days of curing

According to Fig. 1, 5 years after sample preparation, the samples prepared by wet method had lower void ratios, indicating lower settlement. Generally, the behavior of the 1850-day samples was similar to that of the 28-day samples.

According to Fig. 2 which depicts the results of consolidation tests on the samples prepared by dry and wet methods, stabilization with lime resulted in a different behavior than that of the cement-stabilized samples. The samples prepared by dry method had lower void ratios, even though wet- and dry-prepared samples showed very close void ratios and the curve slopes can be observed in both wet- and dry-prepared samples.

According to Fig. 3 where the results of consolidation tests for 1850-day samples stabilized with mixed lime-cement (at equal amounts) by wet/dry method are shown, the samples prepared by dry method had lower void ratios and hence lower settlement levels (compared to the samples prepared by wet method). Another point to note is that, the reduction in the slope of e-log p′ curve can be seen in dry samples with increasing the percentages of stabilizers.

Regarding low variation in void ratios, the variation in permeability of the samples was also low relative to the 28-day samples.

The obtained compressibility coefficient was well similar with that of the 28-day sample, because of small variation in void ratio and standardized loading scheme in the consolidation test. Therefore, the compressibility coefficient was slightly lower than that of the 28-day samples.

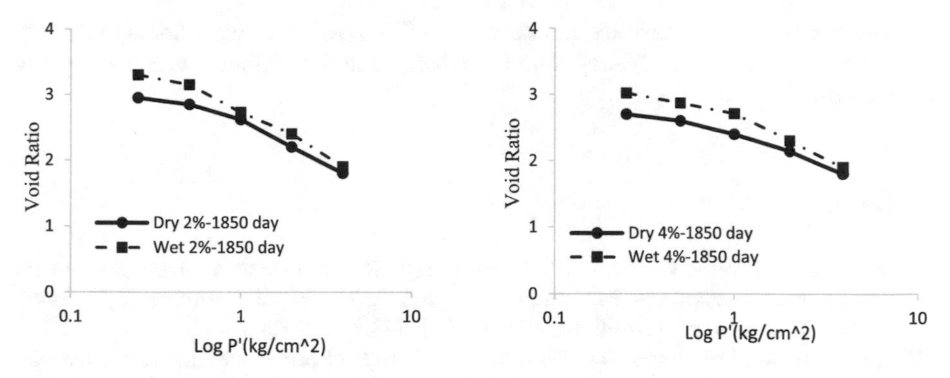

Fig. 2. End of primary e-log p′ relationship for lime treated samples after 1850 days of curing

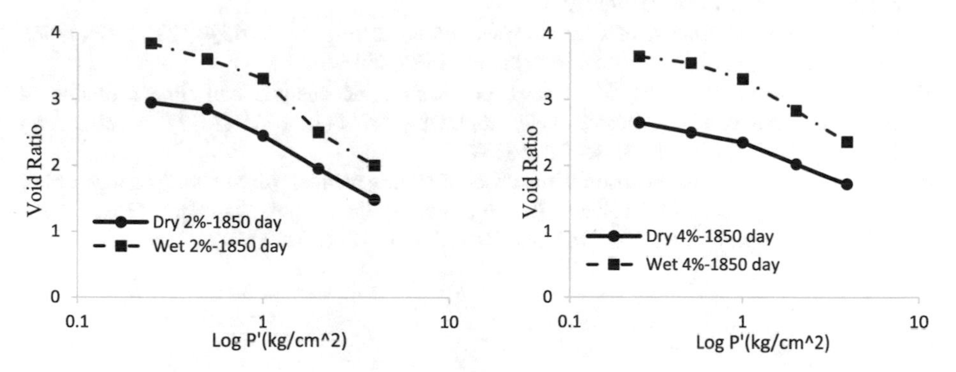

Fig. 3. End of primary e-log p′ relationship for cement-lime treated samples after 1850 days of curing

4 Conclusion

- For all samples, compressive strength followed an increasing trend with the same pattern as that of the 28-day samples.
- Maximum growth was observed in the samples stabilized with 10% of cement by wet method. However, the strength growth rate of the cement-stabilized samples prepared by dry method higher than that of the wet-prepared samples.
- Regarding the results of the consolidation test, the samples containing 6, 8 and 10% of additives showed no significant settlement during the test; therefore, only samples containing 2 and 4% of additives were thoroughly tested by this method.
- The trend of the results of consolidation tests on the samples containing 2 and 4% of additives was similar to that of the 28-day samples, with no significant change observed.
- The cement-containing samples prepared by wet method, rather than dry method, had lower void ratios. However, this pattern was vice versa in the lime and mixed lime-cement-containing samples.

- The changes in permeability and compressibility coefficient were not significantly different from those of 28-day samples, which could be attributed to relatively low variation in void ratio.

References

Nagaraj, H.B., Sravan, M.V., Arun, T.G., Jagadish, K.S.: Role of lime with cement in long-term strength of compressed stabilized earth blocks. Int. J. Sustain. Built Environ. **3**(1), 54–61 (2014). https://doi.org/10.1016/j.ijsbe.2014.03.001. ISSN 2212-6090

Wild, S., Kinuthia, J.M., Jones, G.I., Higgins, D.D.: Effects of partial substitution of lime with ground granulated blast furnace slag (GGBS) on the strength properties of lime-stabilised sulphate-bearing clay soils. Eng. Geol. **51**(1), 37–53 (1998). https://doi.org/10.1016/S0013-7952(98)00039-8. ISSN 0013-7952

Bell, F.G.: Lime stabilization of clay minerals and soils. Eng. Geol. **42**(4), 223–237 (1996). https://doi.org/10.1016/0013-7952(96)00028-2. ISSN 0013-7952

Vitale, E., Deneele, D., Paris, M., Russo, G.: Multi-scale analysis and time evolution of pozzolanic activity of lime treated clays. Appl. Clay Sci. **141**, 36–45 (2017). https://doi.org/10.1016/j.clay.2017.02.013. ISSN 0169-1317

Pakbaz, M.S., Farzi, M.: Comparison of the effect of mixing methods (dry vs. wet) on mechanical and hydraulic properties of treated soil with cement or lime. Appl. Clay Sci. **105–106**, 156–169 (2015). https://doi.org/10.1016/j.clay.2014.11.040. ISSN 0169-1317

A Comprehensive Approach to Determine Intrinsic Compressibility of Reconstituted Clays with Various Initial Water Contents

Jianjun Ma[1], Mingyue Qian[2], Chuang Yu[2(✉)], and Xiaoniu Yu[1(✉)]

[1] College of Architecture and Civil Engineering, Wenzhou University, Chashan University Town, Wenzhou 325035, China
xnyu09@163.com
[2] College of Architecture and Civil Engineering, Wenzhou University, Wenzhou 325035, People's Republic of China
geoyuchuang@163.com

Abstract. Understanding the compressive behavior of reconstituted clays plays the central role for many infrastructures that built in coastal areas. This paper presents a comprehensive approach for the determination of intrinsic compressibility of reconstituted clays with different initial water contents. Following the conventional practice of geotechnical engineering, the compression curve is expressed by two straight lines in the bilogarithmic space, with a remoulded yield stress being introduced to distinguish the two linear segments. On the basis of extensive experimental data from literature, parameters for the description of two linear segments can be obtained through a pair of functions of initial void ratio and void ratio at liquid limit. Then, a comprehensive equation for intrinsic compressibility is formulated to replace empirical approaches based on curve fitting. The proposed equation contains a small number of parameters, which can be obtained through consolidometer tests. This approach is validated through comparison between model predictions and experimental results for remoulded clays under a wide range of effective stresses and initial water contents.

1 Introduction

In coastal cities, many infrastructures (e.g. buildings, roads, railways, tunnels and bridges) are placed on or excavated through reconstituted clays. In engineering design and project assessment, compression curve of reconstituted clay is typically served a reference framework for modelling mechanical behavior of structured soils, owing to the destruction of soil structure (Burland 1990; Cerato and Lutenegger 2004; Hong et al. 2010; Liu and Carter 2000; Najser et al. 2010; Sridharan and Nagaraj 2000; Turchiuli and Fargues 2004). In addition, due to land crisis in coastal cities and storing contaminant soils, reconstituted clay such as dredged soils and sludge have been widely used as a fill for land reclamation (Berilgen et al. 2006; Chu et al. 2012). For extra soft dredged soil and mine tailings, compressibility along with the permeability play significant role in solving engineering problems of waste sludge storage and reclamation (Berilgen et al. 2006; Dolinar 2009; Horpibulsuk et al. 2010; Zeng et al. 2011). Therefore, much more attention has been paid on intrinsic compressibility of

J. S. McCartney and L. R. Hoyos (Eds.): GeoMEast 2018, SUCI, pp. 85–98, 2019.
https://doi.org/10.1007/978-3-030-01914-3_8

reconstituted clays with various water contents (Berilgen et al. 2006; Liu et al. 2013; Zeng and Hong 2016).

Burland (1990) introduced the concept of void index to demonstrate the compression behavior of reconstituted clays in a generalized fashion. This concept was then widely used to compare the mechanical behavior of natural deposit and reconstituted clays (Cotecchia and Chandler 2000; Horpibulsuk et al. 2010; Liu and Carter 2000). However, Burland (1990)'s equation for intrinsic compression line (ICL) was developed for reconstituted clays with initial water contents ranging from 1.0 to 1.5 times the liquid limits, beyond which this equation may be not applicable (Cerato and Lutenegger 2004). Notable contribution on consolidometer tests also demonstrates that initial water content plays a significant role on the compression behavior of reconstituted clays (Berilgen et al. 2006; Cerato and Lutenegger 2004; Hong et al. 2010; Horpibulsuk et al. 2010; Zeng et al. 2016; Zeng and Hong 2015; Zeng et al. 2015). Another limitation of this equation is that the initial effective vertical stress applied by standard oedometer tests is limited to be no less than 6–10 kPa. As noted by Hong (2007), not all soils at their liquid limits are fully virgin, upon compression some fine-grained soils are characterized by 'preconsolidation pressure' or 'remolded yield stress', which may affect Burland's intrinsic compression line significantly. Based on Burland (1990)'s contribution and extensive consolidometer tests on reconstituted clays, Hong et al. (2010) proposed a relationship to extend Burland (1990)'s intrinsic compression line to clays at higher initial water contents and lower effective vertical stresses. Obviously, Hong et al. (2010)'s extended intrinsic compression line is complete and systematic, thus leading the fashion of consolidometer tests later on (Horpibulsuk et al. 2010; Zeng and Hong 2015).

With the extended intrinsic compression line being developed, notable equations has been proposed in literature to describe the compression behavior of reconstituted clays (Hong et al. 2010; Hong et al. 2012; Liu et al. 2013; Zeng et al. 2015). However, most equations are established through curve fitting and model parameters seem to lack clear physical meanings. Consequently, these empirical formulas cannot always predict the compression behavior of reconstituted clays with reasonable accuracy. Moreover, these expressions are formulated based on curve fitting other than rigorous reasoning, no unified relationship for intrinsic compression line can be achieved. Based on Burland's equation and extensive consolidometer tests, this study is going to propose a comprehensive approach to determine intrinsic compressibility of reconstituted clays, with the variation of initial water contents and remolded yield stresses being taken into account.

2 Compression Behavior of Reconstituted Clays

2.1 Experimental Determination of Compression Curve

Extensive oedometer tests show that the compression curve of reconstituted clays is characterized by an inverse 'S' shape in the plot of $e - \log \sigma'_v$ (Cerato and Lutenegger 2004; Hong et al. 2010; Hong et al. 2012; Zeng et al. 2015). For a better interpretation of the oedometer test data, Hong et al. (2010) suggested that this inverse 'S' shape curve can be described by two straight lines in the bilogarithmic space: $\ln(1+e) - \log \sigma'_v$. In this study, as shown in Fig. 1, the compression curve is expressed in the plot of $\ln(1+e) - \log \sigma'_v$. The intersection point of two segments is the remoulded yield stress (σ'_r), with the left line being denoted as the pre-remoulded yield state and the right line as post-remoulded yield state. Following the conventional approach in geotechnical engineering, two sets of model parameters can be applied to describe the proposed two linear segments:

$$\ln(1+e) = \Gamma - \kappa \ln \sigma'_v \quad \sigma'_v \leq \sigma'_r \tag{1}$$

$$\ln(1+e) = \mathrm{N} - \lambda \ln \sigma'_v \quad \sigma'_v \geq \sigma'_r \tag{2}$$

in which, Γ and N represent the logarithmic void ratio $\ln(1+e)$ at intersection points of two straight lines with an initial effective stress of 1 kPa.

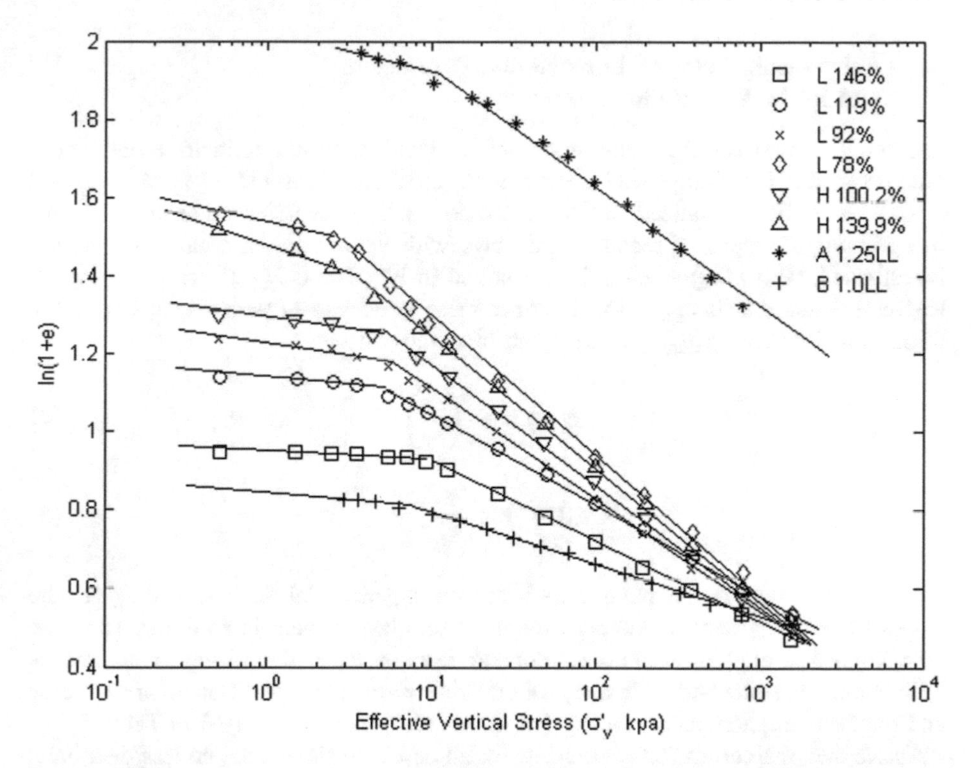

Fig. 1. Compression behaviour of reconstituted clays with various initial water contents: Lianyungang Clays (L, $w_o = 78\% - 146\%$) (Hong et al. 2010), Huainan Clay (H, $w_o = 100.2\% - 139.9\%$) (Zeng et al. 2015), Attapulgite Clay (A, $w_o = 1.25LL$) and Boston Blue Clay (B, $w_o = 1.0LL$) (Cerato and Lutenegger 2004).

Therefore, reconstituted clay with a specific initial water content or void ratio (w_o or e_o) is characterised by its own set of model parameter ($\Gamma(e_o)$, $\kappa(e_o)$, $\mathrm{N}(e_o)$, $\lambda(e_o)$, $\sigma'_r(e_o)$). In terms of the transition state $\sigma'_r(e_o)$, $e_r(e_o)$ at the remoulded yield stress point, Hong et al. (2010) established following equations to describe the relationship between remoulded yield stress, normalized void ratio $e_r(e_o)/e_L$ and normalized initial void ratio e_o/e_L:

$$\sigma'_r(e_o) = 5.66\left(\frac{e_o}{e_L}\right)^{-2} \tag{3a}$$

$$\frac{e_r(e_o)}{e_L} = 2\left[\sigma'_r(e_o)\right]^{-0.42} \tag{3b}$$

Where, $\sigma'_r(e_o)$ and $e_r(e_o)$ are the remoulded yield stress and void rate; e_L is the void ratio at liquid limit state.

2.2 Relationship Between Compression Parameters and Initial Void Ratio

Inspired by the relationship between remoulded yield stress and normalized initial void ratio (Eqs. (3a), (3b)) proposed by Hong et al. (2010), it is reasonable to plot $\kappa(e_o)$ and $\lambda(e_o)$ versus the normalized initial void ratio (e_o/e_L) for some reconstituted soils. In this study, 7 types of reconstituted clays with various initial water contents are investigated, with $\kappa(e_o) - e_o/e_L$ being plotted in Fig. 2 and $\lambda(e_o) - e_o/e_L$ in Fig. 3. Regression analysis is applied to interpret these compression parameters, best curve fitting indicates that, $\kappa(e_o)$ and $\lambda(e_o)$ can be expressed by:

$$\kappa(e_o) = \kappa_L\left(\frac{e_o}{e_L}\right)^{\alpha} \tag{4}$$

$$\lambda(e_o) = \beta \ln\left(\frac{e_o}{e_L}\right) + \lambda_L \tag{5}$$

where, α and β are model parameters based on experimental data, κ_L and λ_L are the slopes for two segments of compression curve for clays at their liquid limits. They can be calibrated through a set of consolidometer tests on reconstituted clays with various initial water contents and wide range of effective vertical stresses. Details of each clay and their model parameters investigated in this study are summarized in Table 1.

Note that, the correlation coefficients for all cases considered are no less than 0.94. Considering the difficulties involved in the consolidometer tests for reconstituted soils with wide range of initial water content and starting from very low effective vertical stresses, and the fact that experimental data from 7 different cases are applied for regression analysis, it is reasonable to state that a strong correlation exists between compression parameters ($\kappa(e_o)$, $\lambda(e_o)$), and initial void ratio.

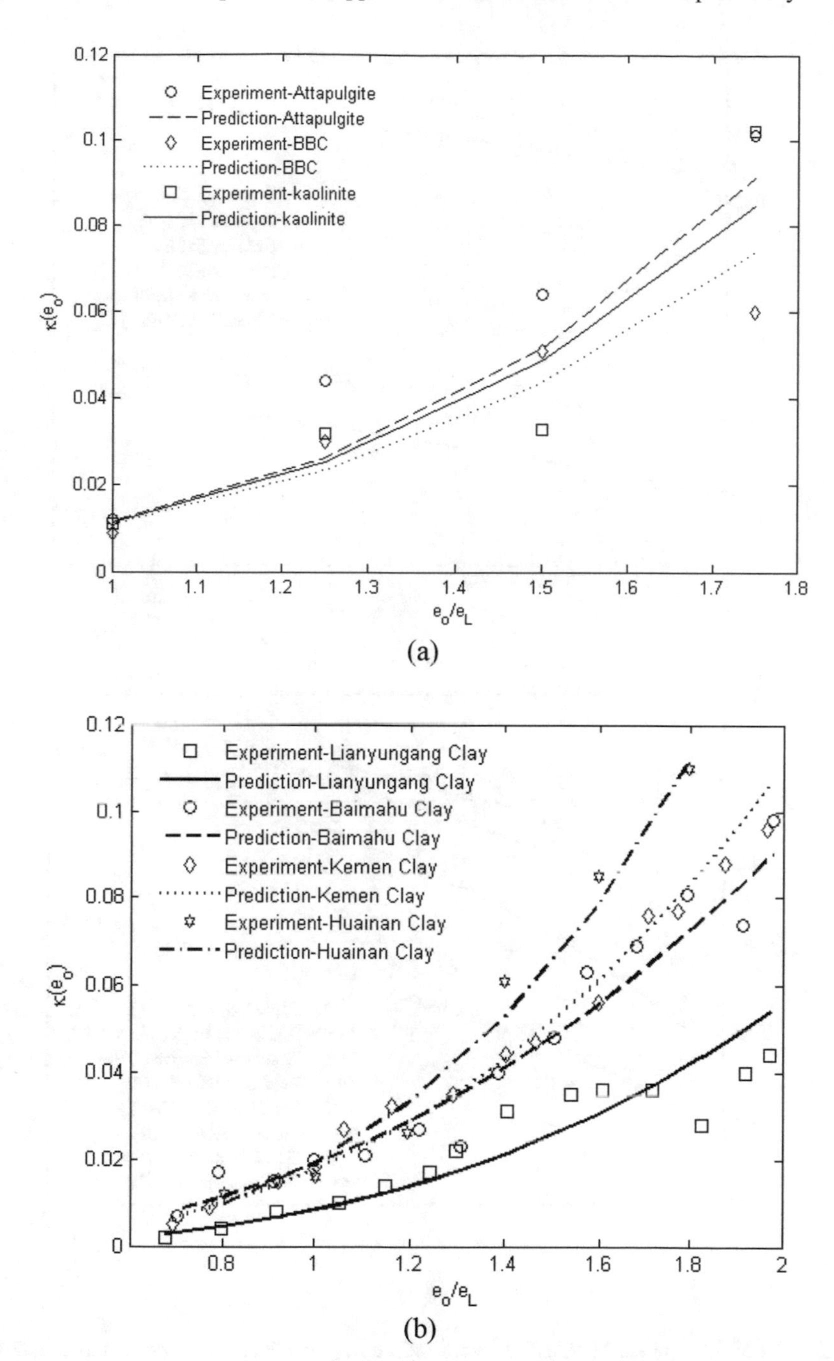

Fig. 2. Plot of $\kappa(e_o)$ against normalised void ratio e_o/e_L: (a) data from Cerato and Lutenegger (2004), (b) data from Hong et al. (2010) and Zeng et al. (2015).

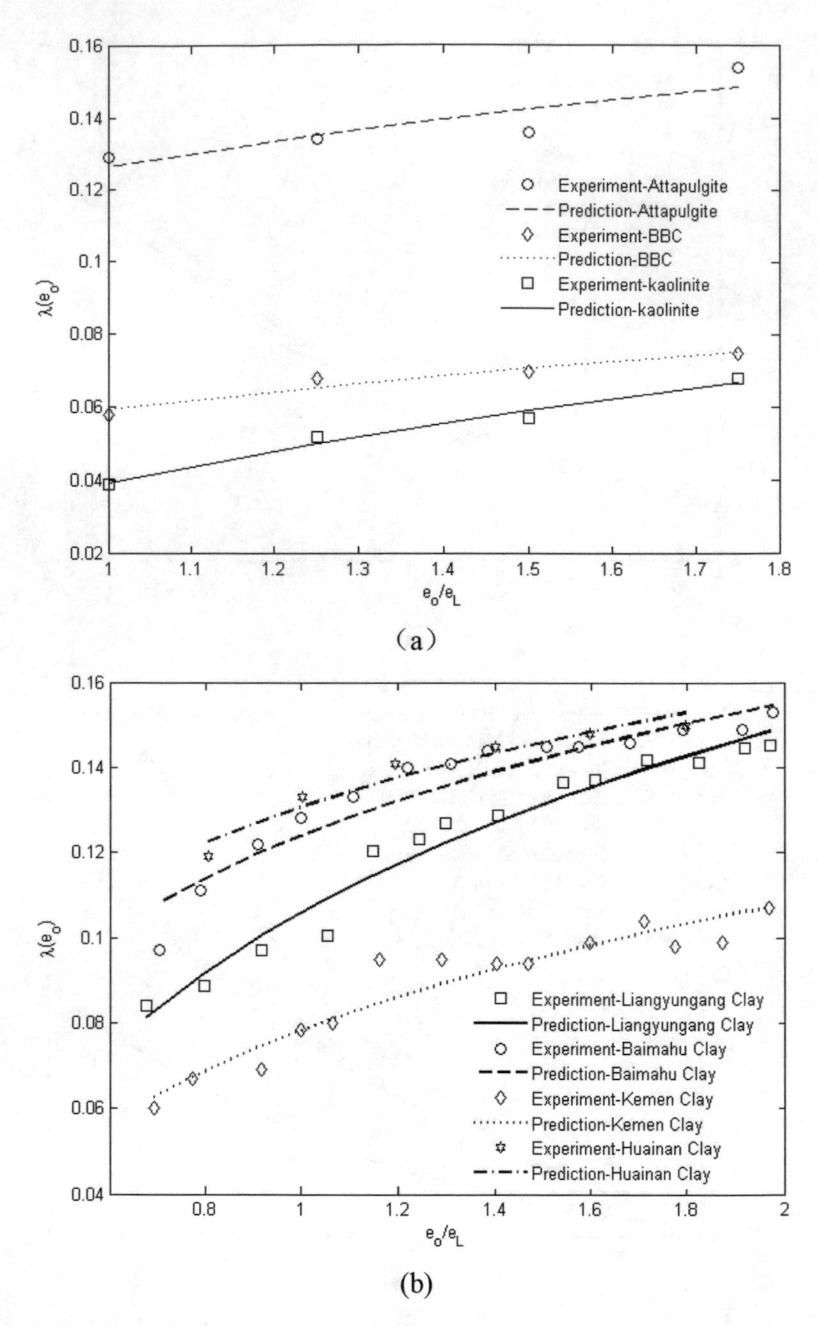

Fig. 3. Plot of $\lambda(e_o)$ against normalised void ratio e_o/e_L: (a) data from Cerato and Lutenegger (2004), (b) data from Hong et al. (2010) and Zeng et al. (2015).

Table 1. Details of reconstituted clays and their model parameters applied for compression curve

Clays	e_o/e_L	κ_L	α	λ_L	β	Data source
Liangyungang clay	0.7–2.0	0.0085	2.72	0.1061	0.0629	Hong et al. (2010)
Baimahu Clay	0.7–2.0	0.0189	2.29	0.1241	0.0472	Hong et al. (2010)
Kemen Clay	0.7–2.0	0.0176	2.65	0.0783	0.0427	Hong et al. (2010)
Huainan Clay	0.806–1.796	0.0191	3.02	0.0377	0.1308	Zeng et al. (2015)
Attapulgite Clay	1.0–1.175	0.0143	3.70	0.0398	0.1264	Cerato and Lutenegger (2004)
Boston Blue Clay (BBC)	1.0–1.175	0.0109	3.42	0.0288	0.0592	Cerato and Lutenegger (2004)
Kaolinite	1.0–1.175	0.0114	3.58	0.0493	0.0393	Cerato and Lutenegger (2004)

3 Relationship for Intrinsic Compressibility

With the expressions of a complete compression curve for reconstituted clays, the void ratio can be expressed as

$$e = \begin{cases} \exp(\Gamma)\left(\sigma_v'\right)^{-\kappa(e_o)} - 1 & \sigma_v' \leq \sigma_r'(e_o) \\ \exp(\mathrm{N})\left(\sigma_v'\right)^{-\lambda(e_o)} - 1 & \sigma_v' \geq \sigma_r'(e_o) \end{cases} \tag{6}$$

Accordingly, the void ratio at 100 kPa e_{100} can take the form

$$e_{100} = \exp(\mathrm{N}(e_o))(100)^{-\lambda(e_o)} - 1 \tag{7}$$

Substituting Eq. (3a), (3b) into (2) and reconsidering the transition state $\left(\sigma_r'(e_o), e_r(e_o)\right)$ at remoulded yield stress $\sigma_r'(e_o)$ yields

$$\exp(\mathrm{N}(e_o)) = \left[1 + 2e_L\left(\frac{5.66}{(e_o/e_L)^2}\right)^{-0.42}\right]\left(\frac{5.66}{(e_o/e_L)^2}\right)^{\lambda(e_o)} \tag{8}$$

thus, e_{100} can be rewritten as

$$e_{100} = \left[1 + 2e_L\left(\frac{5.66}{(e_o/e_L)^2}\right)^{-0.42}\right]\left(\frac{5.66}{(e_o/e_L)^2}\right)^{\lambda(e_o)}(100)^{-\lambda(e_o)} - 1 \tag{9}$$

Intrinsic compression index (c_c) and void index (I_v) can be achieved by their definitions:

$$\begin{aligned} c_c &= e_{100} - e_{1000} \\ &= \left[\exp(\mathrm{N})(100)^{-\lambda(e_o)} - 1\right] - \left[\exp(\mathrm{N})(1000)^{-\lambda(e_o)} - 1\right] \\ &= \exp(\mathrm{N})\left[100^{-\lambda(e_o)} - 1000^{-\lambda(e_o)}\right] \end{aligned} \tag{10}$$

$$\begin{aligned} \mathrm{I}_v &= \frac{e - e_{100}}{e_{100} - e_{1000}} \\ &= \begin{cases} \dfrac{\left[\exp(\Gamma)\left(\sigma'_v\right)^{-\kappa(e_o)} - 1\right] - \left[\exp(\mathrm{N})(100)^{-\lambda(e_o)} - 1\right]}{\left[\exp(\mathrm{N})(100)^{-\lambda(e_o)} - 1\right] - \left[\exp(\mathrm{N})(1000)^{-\lambda(e_o)} - 1\right]} & \sigma'_v \le \sigma'_r \\ \dfrac{\left[\exp(\mathrm{N})\left(\sigma'_v\right)^{-\lambda(e_o)} - 1\right] - \left[\exp(\mathrm{N})(100)^{-\lambda(e_o)} - 1\right]}{\left[\exp(\mathrm{N})(100)^{-\lambda(e_o)} - 1\right] - \left[\exp(\mathrm{N})(1000)^{-\lambda(e_o)} - 1\right]} & \sigma'_v \ge \sigma'_r \end{cases} \\ &= \begin{cases} \dfrac{\exp(\Gamma - \mathrm{N})\left(\sigma'_v\right)^{-\kappa(e_o)} - (100)^{-\lambda(e_o)}}{100^{-\lambda(e_o)} - 1000^{-\lambda(e_o)}} & \sigma'_v \le \sigma'_r \\ \dfrac{\left(\sigma'_v\right)^{-\lambda(e_o)} - 100^{-\lambda(e_o)}}{100^{-\lambda(e_o)} - 1000^{-\lambda(e_o)}} & \sigma'_v \ge \sigma'_r \end{cases} \end{aligned} \tag{11}$$

where, e_{1000} is the void ratio at 1000 kPa. Note that, $\lambda(e_o)$ and $\kappa(e_o)$ can be obtained through the formulations given by Eqs. (4) and (5). With the expression of $\exp(\mathrm{N}(e_o))$ by Eq. (8), the intrinsic compression index c_c can take the form

$$c_c = \left[1 + 2e_L\left(\frac{5.66}{(e_o/e_L)^2}\right)^{-0.42}\right]\left(\frac{5.66}{(e_o/e_L)^2}\right)^{\lambda(e_o)}\left[100^{-\lambda(e_o)} - 1000^{-\lambda(e_o)}\right] \tag{12}$$

Thus, a comprehensive relationship for intrinsic compression index between initial void ratio and void ratio at liquid limit is established through a more rigorous and generalized approach.

Considering that, the void ratio at the remoulded yield stress (e_r) can be expressed as

$$\ln(1 + e_r) = \Gamma - \kappa(e_o)\ln\sigma'_r = \mathrm{N} - \lambda(e_o)\ln\sigma'_r \tag{13}$$

Rearranging Eq. (13),

$$\Gamma - \mathrm{N} = [\kappa(e_o) - \lambda(e_o)]\ln\sigma'_r \tag{14}$$

Substituting Eqs. (3a), (3b) and (14) into (11) yields

$$\mathrm{I}_v = \begin{cases} \dfrac{\left[5.66(e_o/e_L)^{-2}\right]^{\kappa(e_o)-\lambda(e_o)}\left(\sigma'_v\right)^{-\kappa(e_o)} - (100)^{-\lambda(e_o)}}{100^{-\lambda(e_o)} - 1000^{-\lambda(e_o)}} & \sigma'_v \leq \sigma'_r \\ \dfrac{\left(\sigma'_v\right)^{-\lambda(e_o)} - 100^{-\lambda(e_o)}}{100^{-\lambda(e_o)} - 1000^{-\lambda(e_o)}} & \sigma'_v \geq \sigma'_r \end{cases} \quad (15)$$

Therefore, a comprehensive relationship between initial void ratio, void ratio at liquid limit, vertical effective stress and void index can be obtained.

4 Application

Application and reliability of the proposed relationships are demonstrated through comparison between experimental data and model prediction on some reconstituted clays reported in the literature. Experimental data from 60 oedometer tests are analyzed in this study, 4 types of clays are reconstituted with wide range of both initial water content and effective vertical stresses. Details of these tests and model parameters are presented in Table 1.

4.1 Intrinsic Compressibility Index

Hong et al. (2010) reported 42 oedomter tests on Liangyungang Clay, Baimahu Clay and Kemen Clay, with initial water contents ranging from 0.7 to 2.0 times their liquid limits and effective vertical stress starting from 0.5 kPa. For the consideration of generalizing compression behavior, compression parameters (e_{100}, c_c) against normalized void ratio (e_o/e_L) is plotted in Figs. 4 and 5, respectively. Model prediction is represented by lines and data from consolidometer tests by discrete symbols.

Figures 4 and 5 demonstrate that good agreement between model prediction and experimental data has been achieved. The proposed models give reasonable prediction for both e_{100} and intrinsic compressibility c_c of three reconstituted clays. Note that, the empirical equation developed by Burland (1990) cannot capture the variation of intrinsic compression parameters with different initial water contents, which also has been highlighted by Hong et al. (2010).

4.2 Void Index

The developed relationship for void index (I_v) is verified by comparing model predictions with experimental data and predictions by some empirical equations proposed by Burland (1990) and Hong et al. (2010). These empirical equations are:

$$I_v = 2.45 - 1.285 \log \sigma'_v + 0.015\left(\log \sigma'_v\right)^3 \quad \text{Burland (1990)} \quad (16)$$

$$I_v = 3.0 - 1.87 \log \sigma'_v + 0.179\left(\log \sigma'_v\right)^2 \quad \text{Hong etal. (2010)} \quad (17)$$

Three types of reconstituted clays are considered in model verification: Liangyungang Clay by Hong et al. (2010), Boston Blue Clay and Attapulgite Clay by Cerato and Lutenegger (2004), with predicted data being presented by lines and

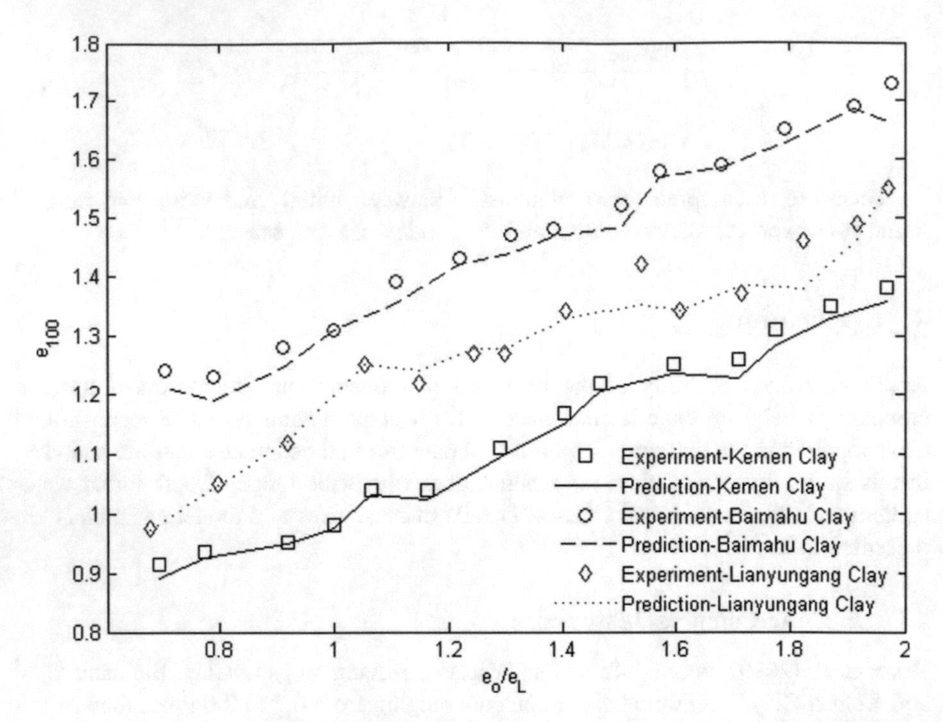

Fig. 4. Comparison between predicted e_{100} and experimental data reported by Hong et al. (2010).

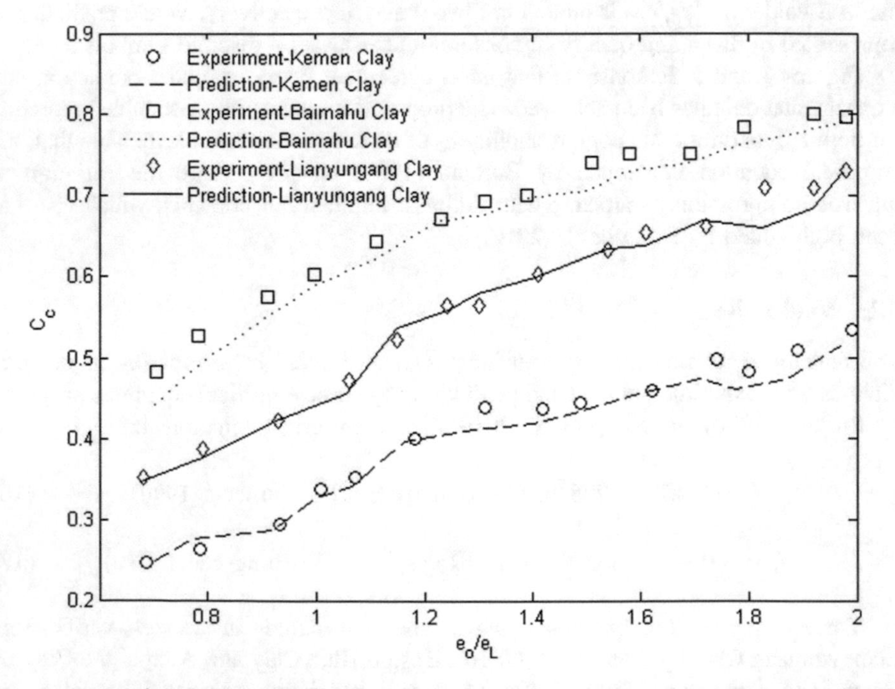

Fig. 5. Comparison between predicted c_c and experimental data reported by Hong et al. (2010)

experimental data by symbols. Boston Blue Clays and Attapulgite Clays were reconstituted with 4 initial water contents: 1.0, 1.25, 1.5 and 1.75 times their liquid limits, the effective vertical stress started from a low value of around 3 kPa.

Inspection of Figs. 6, 7 and 8 demonstrates that, predictions by the proposed comprehensive equation match experimental data well for all cases considered. This model is able to capture the void index curve composed of two linear segments in the $I_v - \log \sigma'_v$ space. The implication of this is that, both the pre-remolded yield state and post-remolded yield state can be described reasonably by the proposed comprehensive equation. Instead, predictions by empirical equations developed by Hong et al. (2010) and Burland (1990) can only match that of some reconstituted clays with a very narrow range of initial water contents. Both of them cannot capture the void index under pre-remolded yield state. Thus, the proposed intrinsic void index is more theoretical and rigorous compared with empirical approaches.

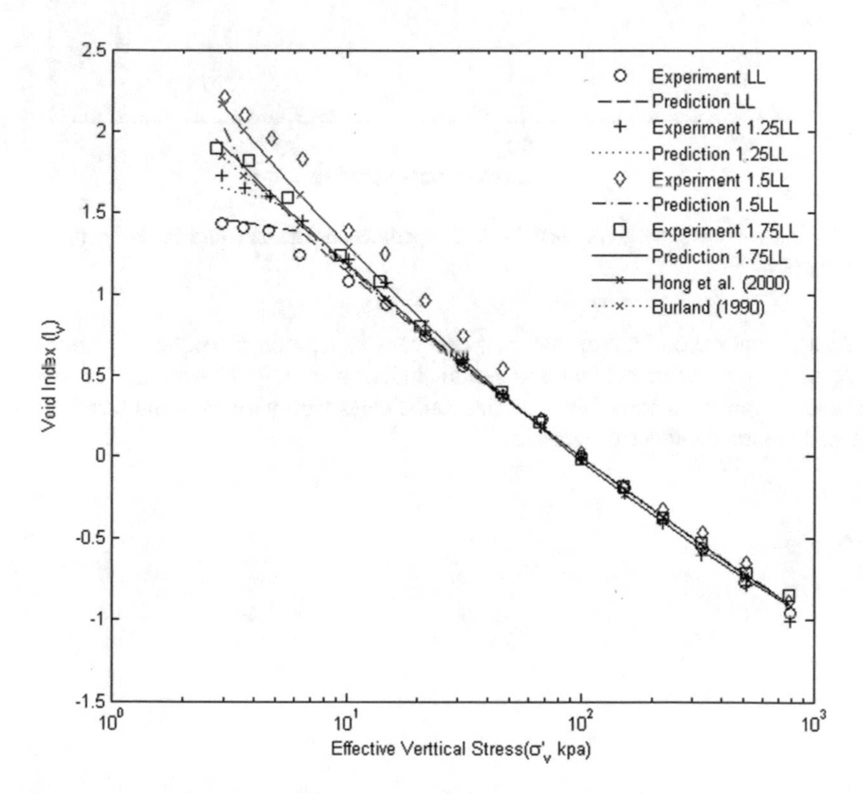

Fig. 6. Comparison between predicted I_v and experimental data on Boston Blue Clay (BBC) (Cerato and Lutenegger 2004).

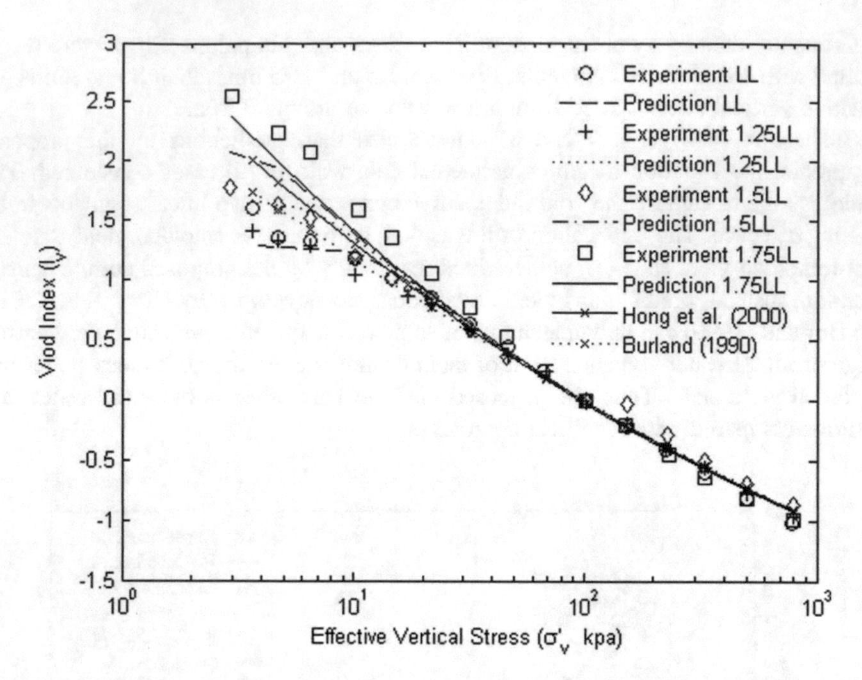

Fig. 7. Comparison between predicted I_v and experimental data on Attapulgite clay (Cerato and Lutenegger 2004).

Model application implies that, a comprehensive relationship between intrinsic void index and initial water content and liquid limit can be a good solution for assessing intrinsic compression behavior of reconstituted clays with various initial water contents and wide range of effective stresses.

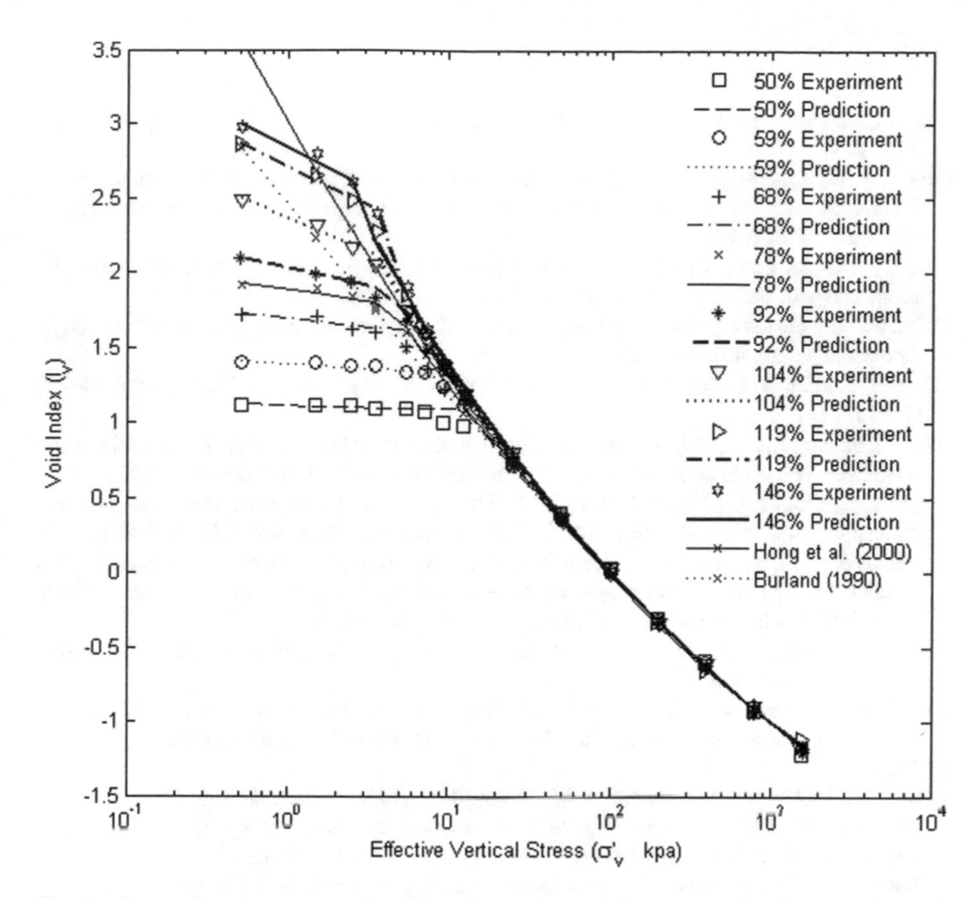

Fig. 8. Comparison between predicted I_v and experimental data reported by Hong et al. (2010)

5 Conclusions

A comprehensive approach for the determination of intrinsic compressibility of reconstituted clays with different initial water contents is proposed. This approach contains a small number of model parameters, which can be calibrated easily through consolidometer tests. The proposed equations are validated through comparison between predictions and experimental results for clays under a wide range of effective stresses and initial water contents. Good performance of the proposed comprehensive relationships demonstrates their capability and reliability to assess intrinsic compressibility of reconstituted clays with various initial water contents.

Acknowledgements. This work is supported by Provincial Commonweal Science Foundation of Zhejiang (PCSFZ, NO.2015C33220, 2017C33220), National Natural Science Foundation of China (NSFC, NO. 51508418, 51508416, 51578427, 41372264). The financial support is gratefully acknowledged.

References

Burland, J.B.: On the compressibility and shear strength of natural clays. Geotechnique **40**, 329–378 (1990)

Cerato, A.B., Lutenegger, A.J.: Determining intrinsic compressibility of fine-grained soils. J. Geotech. Geoenvironmental Eng. **130**, 872–877 (2004). https://doi.org/10.1061/(asce)1090-0241(2004)130:8(872)

Chu, J., Yan, S., Lam, K.P.: Methods for improvement of clay slurry or sewage sludge. In: Proceedings of the ICE - Ground Improvement, vol. 165, pp. 187–199 (2012)

Cotecchia, F., Chandler, R.J.: A general framework for the mechanical behaviour of clays. Geotechnique **50**, 431–447 (2000)

Hong, Z.: Void ratio-suction behavior of remolded ariake clays. Geotech. Test. J. **30**, 234–239 (2007)

Hong, Z.S., Yin, J., Cui, Y.J.: Compression behaviour of reconstituted soils at high initial water contents. Geotechnique **60**, 691–700 (2010). https://doi.org/10.1680/geot.09.P.059

Hong, Z.S., Zeng, L.L., Cui, Y.J., Cai, Y.Q., Lin, C.: Compression behaviour of natural and reconstituted clays. Geotechnique **62**, 291–301 (2012). https://doi.org/10.1680/geot.10.P.046

Horpibulsuk, S., Liu, M.D., Liyanapathirana, D.S., Suebsuk, J.: Behaviour of cemented clay simulated via the theoretical framework of the Structured Cam Clay model. Comput. Geotech. **37**, 1–9 (2010) http://dx.doi.org/10.1016/j.compgeo.2009.06.007

Liu, M.D., Carter, J.P.: Modelling the destructuring of soils during virgin compression. Geotechnique **50**, 479–483 (2000)

Liu, M.D., Zhuang, Z., Horpibulsuk, S.: Estimation of the compression behaviour of reconstituted clays. Eng. Geol. **167**, 84–94 (2013). https://doi.org/10.1016/j.enggeo.2013.10.015

Najser, J., Pooley, E., Springman, S.M., Laue, J., Boháč, J.: Mechanisms controlling the behaviour of double-porosity clay fills; in situ and centrifuge study. Q. J. Eng. Geol. Hydrogeol. **43**, 207–220 (2010). https://doi.org/10.1144/1470-9236/08-033

Sridharan, A., Nagaraj, H.B.: Compressibility behaviour of remoulded, fine-grained soils and correlation with index properties. Can. Geotech. J. **37**, 712–722 (2000)

Turchiuli, C., Fargues, C.: Influence of structural properties of alum and ferric flocs on sludge dewaterability. Chem. Eng. J. **103**, 123–131 (2004). http://dx.doi.org/10.1016/j.cej.2004.05.013

Zeng, L.L., Hong, Z.S., Cui, Y.J.: Determining the virgin compression lines of reconstituted clays at different initial water contents. Can. Geotech. J. **52**, 1408–1415 (2015). https://doi.org/10.1139/cgj-2014-0172

Developing an Experimental Strategy to Investigate Stress-Strain Models Using Kaolin

Mair E. W. Beesley(✉), Paul J. Vardanega, and Erdin Ibraim

Department of Civil Engineering, University of Bristol, Bristol, UK
{mb0126, p.j.vardanega, Erdin.Ibraim}@bristol.ac.uk

Abstract. Prediction of ground movements requires a reliable estimation of soil representative stress-strain behaviour. To do this an assessment of in-situ ('preshear') conditions and the associated influence on the average mobilised soil strength and strain is needed. While many studies focus on undrained shear strength, less effort has been reported for soil shear strain in the context of foundation design. The influence of different experimental and prediction techniques to determine representative soil shear stress-strain design parameters is worthy of study. In this paper, new experimental data is presented of from a series of triaxial and oedometer tests on kaolin. The results demonstrate increasing values of normalised undrained shear strength and reference shear strain with increasing *OCR*.

1 Introduction

When designing geo-structures in fine-grained soils, assessing the average resistance and displacement across a mechanism during undrained shear is challenging due to the uncertain influences of previous stress history and subsequent directions of applied shear stress. If a representative stress-strain curve is required, a triaxial test can be used to mimic the in-situ stress and drainage conditions during shear (e.g. Bishop and Henkel 1957). Reconsolidating anisotropically to the in-situ stresses arguably yields a better estimate of in-situ soil parameters (Bjerrum 1973); however, isotropic consolidation is more commonly encountered in practice since the procedure is simpler and less expensive. Bjerrum (1973) recommended that samples should be tested in the laboratory under a variety of test modes (Triaxial compression, UC, Triaxial extension, UE, Direct Simple Shear, DSS) and the undrained shear strengths (c_u) obtained compared to those from field vane tests. Many researchers have investigated the variation in c_u with stress history and direction (Jamiolkowski et al. 1985; Ladd and Foott 1974) but little guidance on similar assessments for soil strain is available.

J. S. McCartney and L. R. Hoyos (Eds.): GeoMEast 2018, SUCI, pp. 99–118, 2019.
https://doi.org/10.1007/978-3-030-01914-3_9

2 Research Objectives

In this paper, new experimental data of reconstituted kaolin samples which have been tested under various conditions of consolidation and undrained shear are presented. The influence of varying initial void ratio and consolidation rate on the derived compressibility parameters (λ and κ) in one-dimensional (K_0) and isotropic conditions is investigated. In addition, the bilogarithmic representation (Butterfield 1979) of the compression data is used to derive analogous λ^* and κ^* parameters. The sensitivity of measured stress-strain behaviour and strength to different undrained shearing conditions is examined using alternative representations of the experimental data. The values of strength and strain parameters from each sheared sample are shown to be affected by the assumed shape of sample deformation.

For foundation design, engineers may need to predict the variation in shear strain and c_u for a range of overconsolidation ratio (OCR). To model the changes in c_u due to changing OCR the formulation based on that presented in Ladd et al. (1977) (Eq. 1) is used in this paper. Mayne (1980) compiled a comprehensive database of experimental evidence which supports the validity of Eq. 1 for a range of soils.

$$\frac{\left(\frac{c_u}{p_0'}\right)_{OC}}{\left(\frac{c_u}{p_0'}\right)_{NC}} = OCR^{\wedge} \tag{1}$$

Strength mobilisation of fine-grained materials may be characterised using a simple power law (Vardanega and Bolton 2011, Vardanega et al. 2012) (Eq. 2)

$$\frac{1}{M} = \frac{\tau_{mob}}{c_u} = 0.5\left(\frac{\gamma}{\gamma_{M=2}}\right)^b \quad 0.2 < \tau_{mob}/c_u < 0.8 \tag{2}$$

For serviceability calculations, Eq. (2) can be used to predict shear strains where $\gamma_{M=2}$ and b are model parameters calculated from shear stress-strain data (Vardanega and Bolton 2011). Vardanega et al. (2012) presented a correlation (Eq. 3) between $\gamma_{M=2}$ and OCR using overconsolidated kaolin samples sheared in triaxial compression:

$$\gamma_{M=2} = 0.0040(OCR)^{0.680} \quad n = 18, R^2 = 0.815 \tag{3}$$

3 Method

The kaolin material used in this study was obtained from two different suppliers, identified as Batch 1 and Batch 2 in this research. Table 1 shows the mean and range of measured Atterberg limits and specific gravity for each batch of kaolin. Liquid limit (w_L) was measured using the fall cone penetrometer and the thread-rolling test was used to measure plastic limit (w_P) as per the requirements given in BSI (1990). Specific gravity (Gs) was measured using the standard pyknometer method, following the procedure specified in BSI (1990).

Table 1. Classification test results

KAOLIN		w_L (%)			w_P (%)			G_S		
Reference	Relevant tests	Mean	n	Range	Mean	n	Range	Mean	n	Range
This study Batch1	O-1.35 O-1.50	66.8	4	0.2	35.1	2	0.5	n/a (2.60 assumed)		
This study Batch2	All other tests	65.5	1	–	33.2	2	0.2	2.60	2	0.01
Cerato and Lutenegger (2004)	Oedometer	42	–	–	26	–	–	2.68	–	–
Vardanega et al. (2012)	CIU	62.6	1	–	29.6	4	–			

3.1 Sampling Procedure

The sampling procedure for kaolin triaxial samples was developed from the method described by Bialowas (2016). For all triaxial and oedometer tests, powdered kaolin was initially oven-dried for approximately 12 h which was subsequently cooled for 3 to 4 h. This was followed by hand mixing into a slurry at the water content of 130% before curing overnight. Air bubbles were removed from the cured slurry by applying vacuum seated on a vibrating table typically over a period of 2 h.

For the triaxial samples, the de-aired slurry was then poured into a 50 mm-diameter consolidometer and compressed with three increments of vertical pressure to reach the maximum sampling consolidation stress. 'CIUC-1-b-200' and 'CIUC-1-b-404' were further 'swelled' to $OCR = 2$ in the consolidometer prior to extrusion (although negligible height change was observed during the swelling stage, which was most likely due to large shaft friction mobilised in the tall consolidometer). The extruded sample height was designed to be 100 mm for each triaxial test (a range of 96–102 mm could be achieved). For oedometer samples, the de-aired slurry was poured directly into the oedometer ring.

3.2 Triaxial Test Procedure

Table 2 presents the experimental details of the seven triaxial tests presented in this paper. Two linear voltage displacement transducers (LVDTs) were attached to the top and base caps (Fig. 1) to measure axial strain during consolidation and shear and, in particular, to monitor the change in distance between the attached bender elements reliably. For brevity, the bender test results are not discussed in this paper. Owing to the low strength of the reconstituted clay samples, mid-height LVDTs attached directly to the middle section of the sample were judged to be inappropriate. The local strain measurements will be affected by any bedding of the caps; however, for fine-grained materials, the bedding error is likely to be low (Sarsby et al. 1980).

To achieve saturation, the cell and back pressures were raised simultaneously at a rate of 25 to 50 kPa/hour. Owing to the uncertain effect on p' during extrusion, which could not be measured from apparently inconsistent measurements of residual pore pressure, different values of effective stress were applied during saturation for each test.

Table 2. Triaxial test details - isotropic consolidation and undrained shear –

Triaxial Test:	CI_1	CIUC-1-a-395	CIUC-2-a-208	CIUC-8-a-51	CIUE-8-a-52	CIUC-1-b-200	CIUC-1-b-403	CIUC-2-b-200
Max. applied stress in sampling device (kPa)	60	60	60	60	60	120	120	120
Extruded w_x (%)	66.0	63.1	59.8	64.1	62.5	54.2	52.7	54.3
Saturation								
Duration (days)	2.18	1.06	1.20	2.15	1.24	0.75	1.01	1.13
p'_{sat} (kPa)	6	4	20	7.5	5	35	34	60
Δe_{SAT}	0.012	0.069	−0.068	0.089	0.290	−0.012	0.014	−0.038
$\varepsilon_{a\ \text{SAT}}$ (%)	1.410*	−0.053	−0.010	−0.005	−0.187	−0.203	0.313	2.137[a]
B value	0.947	0.949	0.952	0.959	0.971	0.958	0.947	0.914
Consolidation								
Loading type	Discrete	8 kPa/h	8 kPa/h	8 kPa/h	8 kPa/h	5 kPa/h	5 kPa/h	5 kPa/h
Increments	6	1	3	2	2	4	7	11
Duration (days)	10.21	5.89	3.45	2.95	2.76	8.08	4.88	4.69
Swelling								
Duration (days)	0	0	1.10	4.01	6.13	0	0	10.08
Preshear held stress [b]								
Duration (days)	3.01	0.73	0.05	2.19	4.30	6.19	1.23	7.42
Δe_{PRE}	−0.004*	−.014	0.000	+.006	−.047	−.007	−.008	+.005
e_m	1.142*	1.195	1.158	1.155	1.274	1.236	1.109	1.142
e_0	1.142*	1.195	1.180	1.254	1.271	1.236	1.109	1.159
p'_m (kPa)	395.69	395.16	401.80	399.35	399.85	200.41	403.15	399.87
p'_0 (kPa)	395.47	395.16	208.04	51.30	51.73	200.12	403.15	200.08
OCR	1.0	1.0	1.9	7.8	7.7	1.0	1.0	2.0
Shear mode	Excluded	UC	UC	UC	UE	UC	UC	UC
Load cap		Flat	Flat	Flat	Vacuum-1	Flat	Vacuum-2	Concave
Filter strips		Yes	Yes	Yes	No	No	No	No

Notes: All samples from Kaolin Batch 2; all sheared samples tested with displacement rate = 0.002 mm/minute.
[a]The sample came into contact with the load cell during setup
[b]Duration of effective stress held prior to shear; this is included in the consolidation or swelling durations
*Estimated values assuming an extruded sample height of 100 mm

The changes in void ratio and axial strain during the saturation period are included in Table 2. With the exception of sample 'CIUC-2-b-200', minimum B values of 0.947 were achieved (Skempton 1954 defines the B value).

Samples were isotropically-consolidated using two continuous stress-controlled loading rates of (a) 8 kPa/hour, and (b) 5 kPa/hour. Identifiers (a) and (b) are included as part of each test reference, shown in Table 2. In addition, one isotropic consolidation test ('CI-1') was carried out using discrete increments of total stress with 24-hour drainage intervals. All samples were sheared undrained using a conventional displacement-control frame at a rate of 0.002 mm/minute. Assessment of membrane restraint was conducted according to the method of Lade (2016): the order of magnitude of stress contribution is between 1 and 5%. However, the original values of stress are presented in this paper.

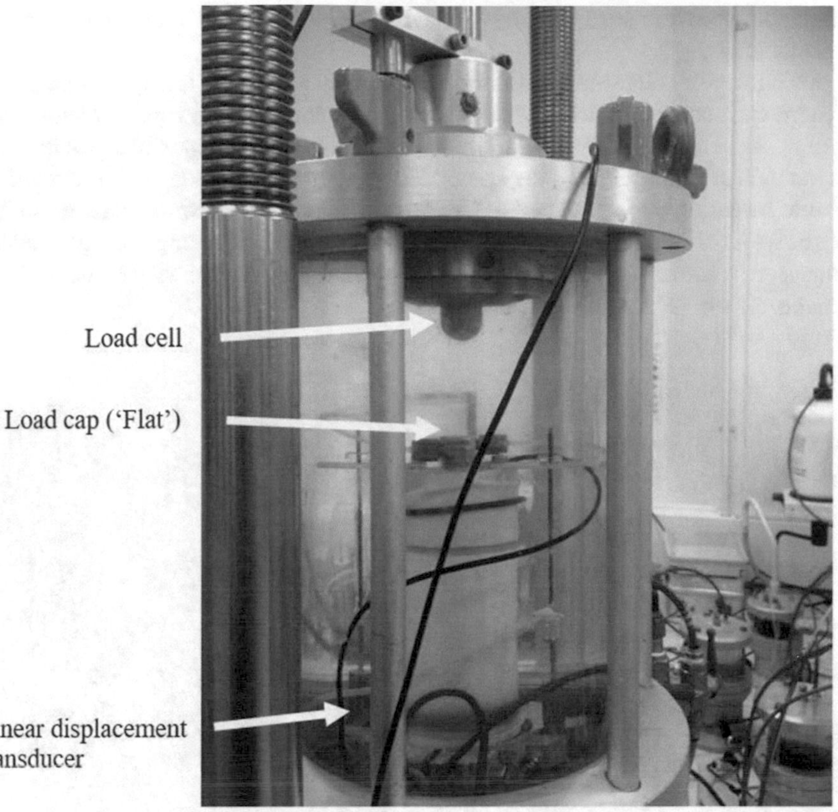

Fig. 1. Compression test setup of kaolin inside conventional isotropic triaxial cell (Linear displacement transducers are attached to the base and top caps with a lightweight connection system)

3.3 Effects of Sample Tilt

The triaxial samples were observed to develop varying degrees of tilt during isotropic consolidation. This led to 2 potential concerns: (1) misalignment of the LVDTs, resulting in loss of local strain data and, more pertinently, in the restraint on sample deformation if the LVDT rods became stuck; (2) misalignment of the load cap connection prior to shear.

To alleviate some of the effects of (1) and (2), the sampling procedure was adapted to produce samples with a lower initial void ratio, with the aim of reducing volumetric compression before undrained shear. To reduce the risk of interference between the LVDTs and sample deformation during the test, the original LVDT connection system was replaced with smooth, lightweight components and a wider range of movement. 'Flat' load cap connections (see the descriptions in Table 2) were used for most tests and a photograph of this connection type is shown in Fig. 1. Any tilt causes some eccentricity of the axial load applied onto the flat surface. A concave surface was manufactured into the top cap for 'CIUC-2-b-200' to observe the difference in behaviour when the load cap was forced into vertical axial alignment when commencing shear.

To undertake extension tests in the conventional triaxial cell, two cap designs using vacuum chambers were investigated: 'Vacuum-1' consists of a rigid vacuum chamber fixed to the internal load cell, which connects to a smooth Perspex plate on the top cap via vacuum seal (see Fig. 2); the second iteration of this component is the rotationally flexible extension cap ('Vacuum-2'), that was designed to accommodate up to 12° of tilt (however, 'CIUC-1-b-403' was tested in compression using this particular cap because a vacuum seal could not be achieved for extension shear due to leakage through the cap joints).

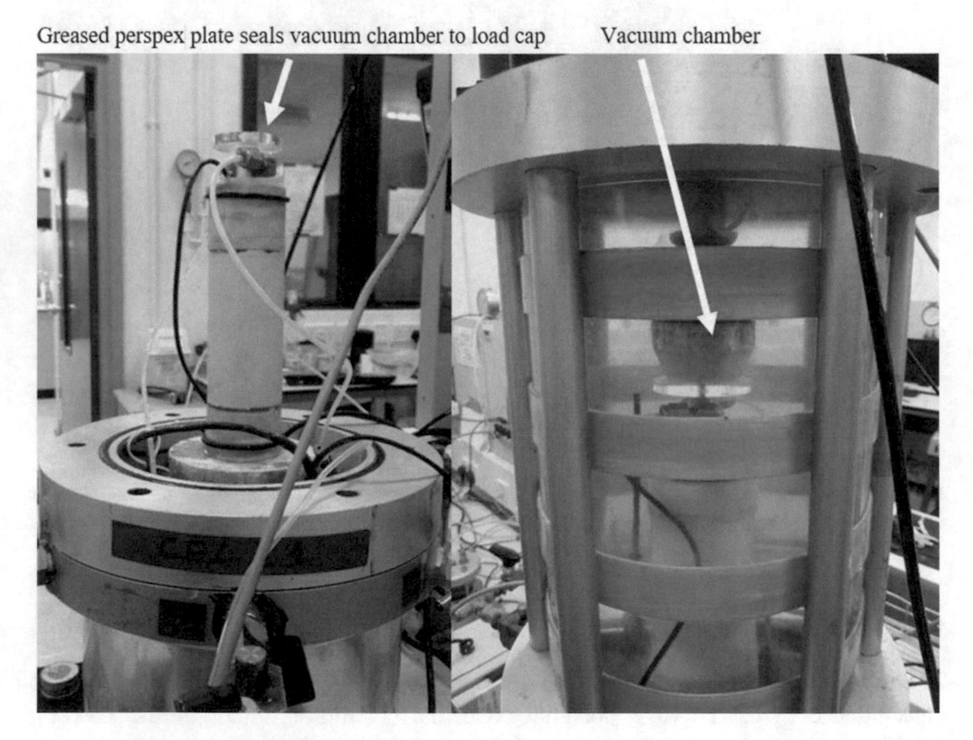

Fig. 2. Extension test setup of kaolin sample CIUE-8-1 inside conventional isotropic triaxial cell with extension cap "Vacuum-1"

3.4 Oedometer Testing

Table 3 presents the details of the three oedometer tests undertaken for this study. For tests 'O-1.35' and 'O-1.50', the conventional oedometer was used i.e. a one-dimensional consolidation device with fixed o-ring. Fixed increments of load were applied every 24 h and vertical displacements were monitored with LVDTs. Test 'O-2.10' was undertaken using a similar setup with the load applied via a stress-controlled triaxial frame.

Table 3. Oedometer test details

Oedometer test ID	Batch no.	w_x %	kPa	σ'_{v0} kPa	No. increments	Loading type	Duration consolidation days	Duration swelling days	Duration total days
O-1.35	1	90.0	400	20	9	Discrete	6	3	9
O-1.50	1	99.2	400	20	9	Discrete	6	3	9
O-2.10	2	138.2	2000	100	20	Discrete	8	6	14
Cerato and Lutenegger (2004)									
*O-1.00		42.0	786	786	15	Discrete	15	0	15
*O-1.25		52.5	778	778	15	Discrete	15	0	15
*O-1.50		63.0	786	786	15	Discrete	15	0	15
*O-1.75		73.5	786	786	15	Discrete	15	0	15

*Tests from the literature included for comparison

4 Experimental Results

4.1 Compressibility of Isotropic Samples

Figure 3 shows the compression data plotted in semi-logarithmic form of the samples undergoing isotropic consolidation. The test data exhibit a pronounced curve in each test, which indicates that every sample had swelled during saturation to a lower effective stress than it had previously experienced. As expected, the apparent preconsolidation stress observed in Fig. 3 is somewhat lower than the stress applied in the consolidometer. For example, Sukolrat (2007) measured in reconstituted Bothkennar samples a loss in applied stress of 35–55% on account of consolidometer side friction.

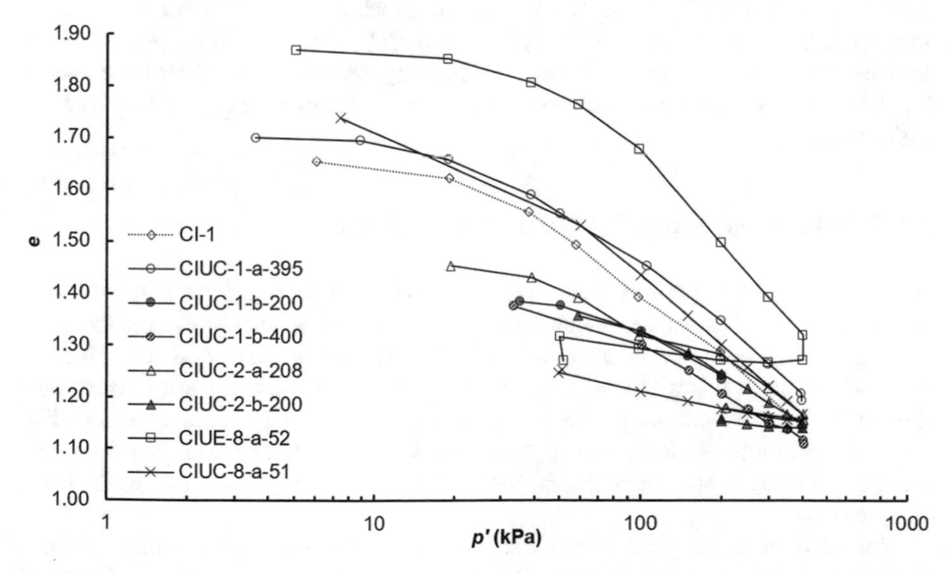

Fig. 3. Semi-logarithmic isotropic-consolidation curves of reconstituted kaolin at different initial water content due to swelling during saturation

From the data presented in Fig. 3, it is possible to distinguish samples which had been preconsolidated at higher stresses in the consolidometer ('CIUC-1-b-200', 'CIUC-1-b-403', and 'CIUC-2-b-200').

The results displayed in Fig. 3 suggest that initially identical kaolin samples will not converge to the same normal consolidation line in the stress range studied. The lack of convergence may be due to differences in swelling during saturation. The value of λ shown in Table 4, obtained by linear regression for each test for stresses generally greater than 200 kPa, varies from 0.124 to 0.258 and increases with the value of void ratio measured at the end of saturation. As expected, the same pattern emerges with values of λ^* (obtained from Fig. 4), although a smaller range is observed.

There appears to be no clear trend in swelling behaviour of the four overconsolidated samples (*OCR* = 2 and 8). Values of κ vary from 0.019 to 0.045 and all samples underwent considerable volume change. The observed changes in void ratio in samples 'CIUC-2-a-208' and 'CIUC-2-b-403' are similar over the same stress range (400 to 200 kPa) despite having different swelling rates. However, the swelling lines for the samples overconsolidated using the same procedure to *OCR* = 8 are markedly different; in particular, 'CIUE-8-a-52' appears to continue to consolidate (or leak) at the end of the swelling period.

4.2 Compressibility of K_0 (oedometer) Samples

Figure 5 shows the compression behaviour of the oedometer samples represented in semi-logarithmic form. The samples tested at higher initial water content exhibit a slightly upward concave shape, indicated by the differences in gradient between the dashed and full lines (corresponding values of λ are provided in Table 4). Since the samples were compressed in the oedometer directly from a slurry, no 'preconsolidation' stress is observed. Values of λ for stresses greater than 60 kPa are very similar, varying from 0.235 to 0.247 with high R^2 and *RD* values (Table 4). Although the number of oedometer tests is limited, the data suggests that K_0-swelling behaviour is dependent on maximum consolidation stress i.e. κ increases with σ'_{vm} and varies from 0.037 to 0.070.

5 Behaviour of Samples Undergoing Shear

Figures 7, 8, 9 and 10 present the behaviour of triaxial samples during strain-controlled undrained shear. From observations of each sample during the shearing stage, little bulging/necking occurred in all cases. The data shown in Figs. 7, 8 and 10 were therefore analysed using the assumption of right cylinder deformation (Bishop and Henkel 1957). The resulting parameters are included in Table 5 and the corresponding values for parabolic (bulging/necking) deformation are shown for comparison. Local strain measurements were used for 0–0.5% axial strain in the stress-strain analysis for every test.

The effective stress paths of the kaolin samples, shown in Fig. 7, follow patterns reasonably consistent with isotropically-consolidated samples at different values of *OCR* (see Wroth and Loudon 1967 for a similar study). The normalised excess pore

Table 4. Calculated compressibility parameters

Test ID	Stress range kPa	n	λ	R^2	RD	λ*	R^2	RD	Stress range kPa	n	κ	R^2	RD	κ *	R^2	RD
Isotropic consolidation																
CI-1	200–400	3	0.208	.999	3.2	0.094	.999	3.2								
CIUC-1-a-395	200–400	3	0.206	1.00	0	0.093	1.00	0								
CIUC-2-a-208	200–400	3	0.164	.999	3.2	0.072	.999	3.2	400–200	3	0.034	.990	10.0	0.015	.990	10.0
CIUC-8-a-51	200–400	5	0.195	1.00	0	0.083	1.00	0	400–51	8	0.045	.993	8.19	0.020	.994	7.8
CIUE-8-a-52	200–400	3	0.258	1.00	0	0.105	1.00	0	400–52	5	0.023	.892	32.9	0.010	.892	32.9
CIUC-1-b-200	100–200	3	0.124	1.00	0	0.052	.999	3.2								
CIUC-1-b-403	200–400	5	0.127	.990	10.0	0.058	.990	10.0								
CIUC-2-b-200	200–400	5	0.141	.999	3.2	0.062	.999	3.2	400–200	5	0.019	.927	27.0	0.009	.928	26.8
Vardanega et al. (2012)																
*Average of 4 tests			0.250								0.039					
K_0 (oedometer) consolidation																
O-1.35	10–60	4	0.237	.998	4.6	0.084	.997	5.9								
	60–200	3	0.247	.997	5.8	0.100	.998	4.2	200–20	4	0.037	.981	13.6	0.016	.983	13.2
O-1.50	10–60	4	0.272	.998	4.7	0.094	.999	2.8								
	60–200	3	0.249	.997	5.7	0.099	.998	4.1	200–20	4	0.039	.989	10.6	0.016	.990	10.1
O- 2.10	10–60	3	0.339	.998	4.0	0.114	.999	1.0	100/400	5	0.057	.935	25.6	0.025	.932	26.0
	60–2000	8	0.231	.999	1.7	0.105	.998	4.6	2000–100	7	0.070	.997	5.7	0.036	.995	7.1
Cerato and Lutenegger (2004)																
*O-1.00	153–786	5	0.069	0.998	4.5	0.036	.998	4.1								
*O-1.25	153–778	5	0.079	0.998	4.5	0.042	.999	3.2								
*O-1.50	153–786	5	0.088	0.997	5.5	0.045	.998	4.5								
*O-1.75	152–786	5	0.091	0.999	3.2	0.046	1.0	0								

*Tests from the literature included for comparison

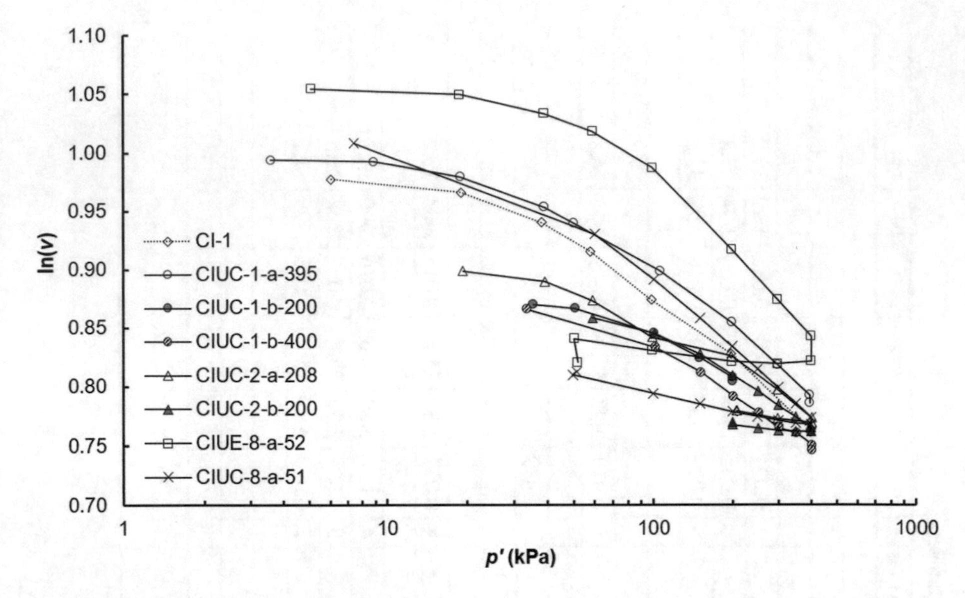

Fig. 4. Bilogarithmic isotropic-consolidation curves of reconstituted kaolin at different initial water content due to swelling during saturation

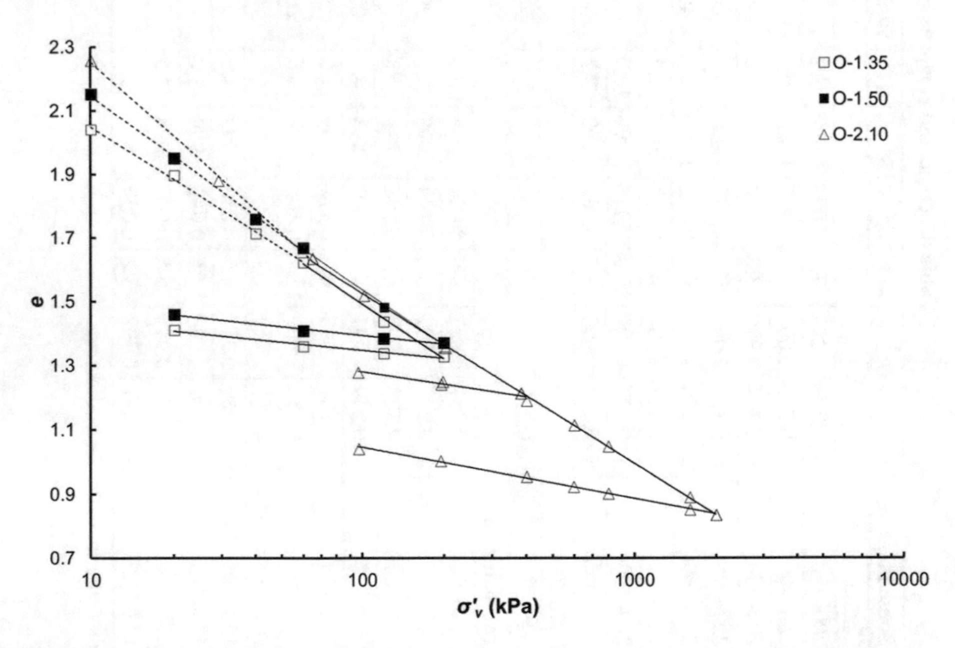

Fig. 5. Semi-logarithmic K_0-consolidation (oedometer) curves of reconstituted kaolin mixed at different initial water content

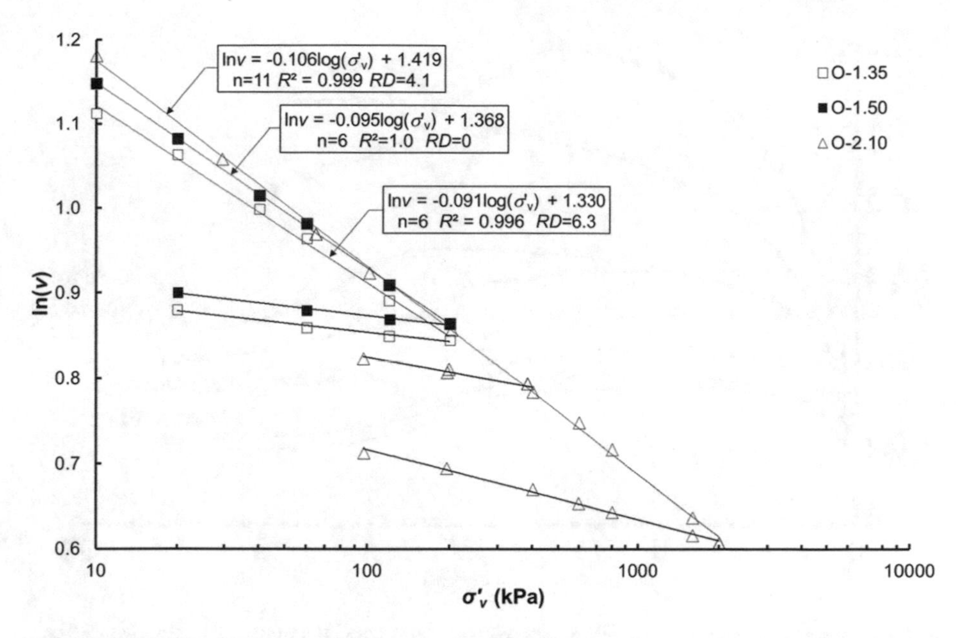

Fig. 6. Bilogarithmic K_0-consolidation (oedometer) curves of reconstituted kaolin mixed at different initial water content

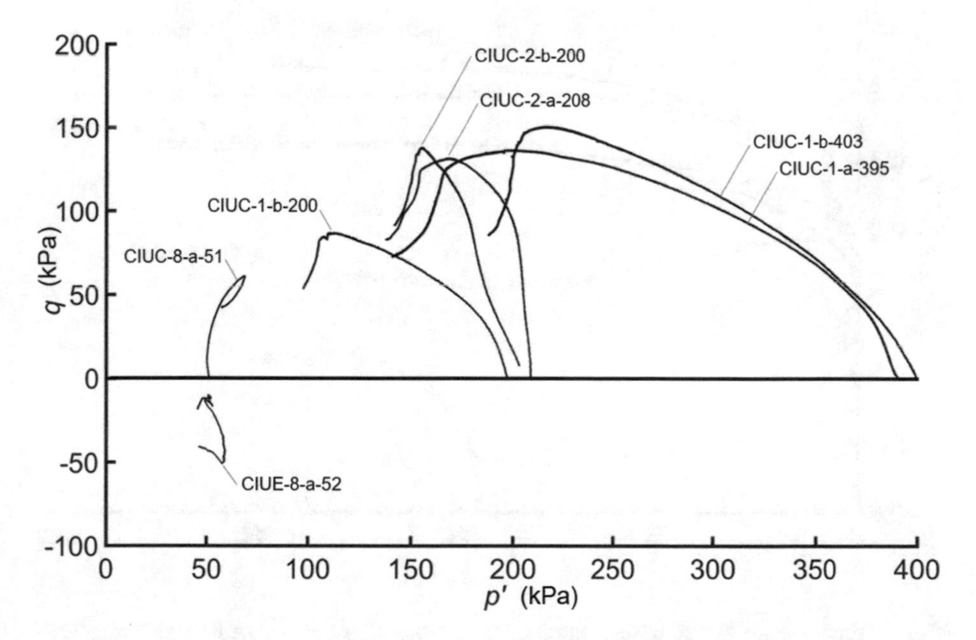

Fig. 7. Effective stress paths for isotropically consolidated triaxial samples

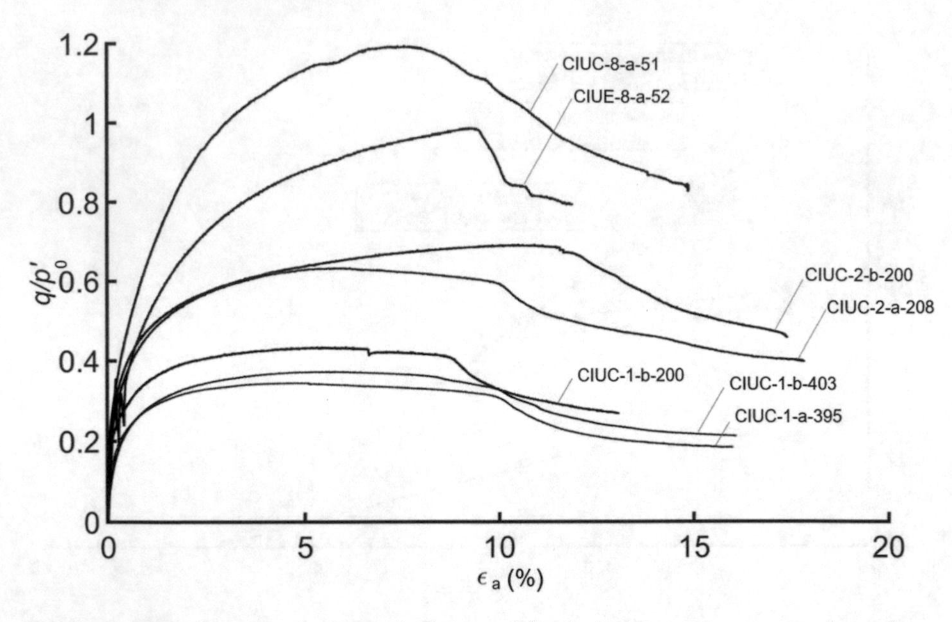

Fig. 8. Stress-strain curves for isotropically consolidated triaxial samples – comparison of tests by p'_0 normalisation (assumed cylinder deformation)

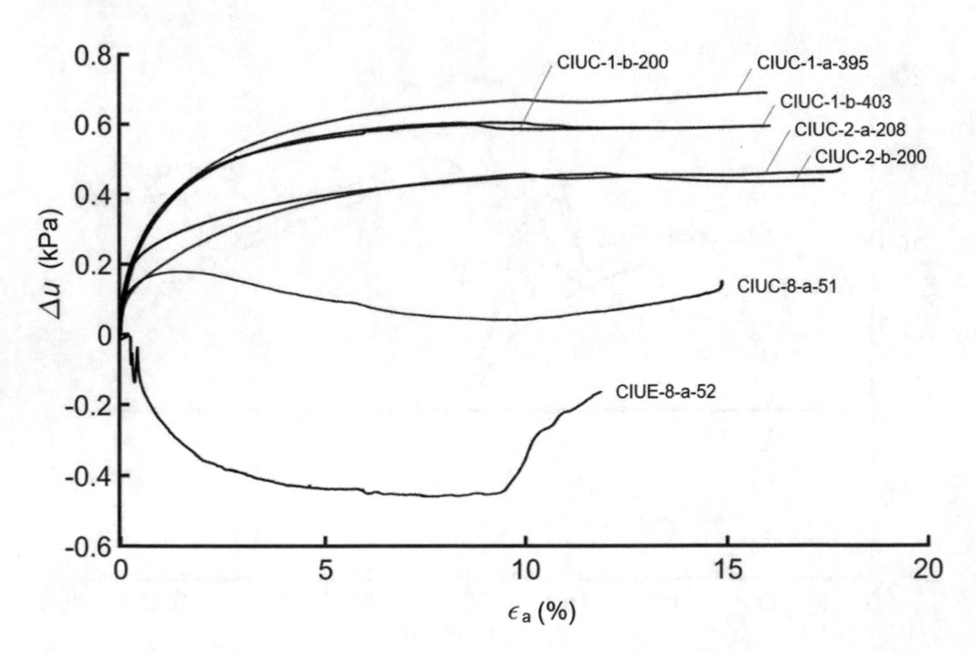

Fig. 9. Excess pore pressure-strain curves for isotropically consolidated triaxial samples – comparison of tests by p'_0 normalisation

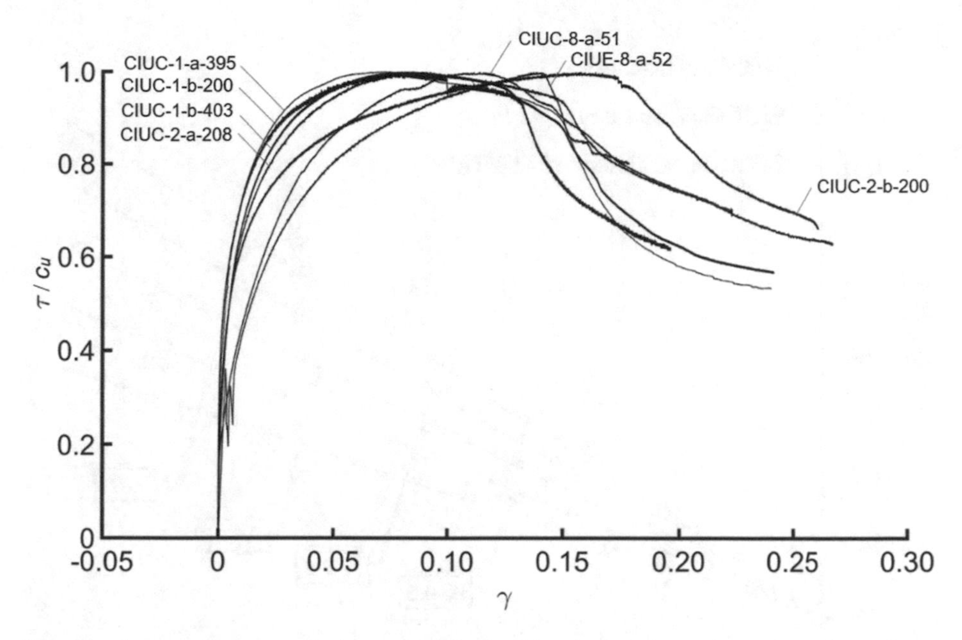

Fig. 10. Stress-strain curves for isotropically consolidated triaxial samples – comparison of tests by c_u normalisation (assumed cylinder deformation)

pressures generated in the normally consolidated samples (Fig. 9) exceed those measured in the lightly overconsolidated (*OCR* = 2) samples, resulting in a more pronounced curve in effective stress path prior to peak failure. Stress paths for pairs of samples consolidated in a similar manner (CIUC-1-a-395, CIUC-1-b-403, and CIUC-2-a-208, CIUC-2-b-200) are close in shape although some differences are observed in mean effective stress and peak deviator stress. The effective stress behaviour of the overconsolidated (*OCR* = 8) samples correspond with the development of negative excess pore pressures; although the sample sheared in extension appears to show an increased tendency to dilate.

The data in Fig. 8 illustrate the effect of overconsolidation on the stress-strain behaviour and peak undrained shear strength, with higher values of deviator stress reached at higher values of *OCR*. Additionally, the increments in stress measured up to around $\varepsilon_a = 4\%$ are very similar for each pair of samples consolidated in a similar manner (CIUC-1-a-395, CIUC-1-b-403, CIUC-2-a-208, and CIUC-2-b-200). Figure 10 highlights the differences in mobilised strain between tests: if CIUC-2-b-200 is considered to be anomalous (due to substantial bedding which occurred at the start of shear), it appears that strain to failure increases with increasing *OCR*. The influence of overconsolidation on mobilised strain appears to be dominant at high values of stress ratio ($\tau_{mob}/c_u > 0.5$).

Figure 11 shows that the strength ratio (taken as the maximum value of deviator stress normalised by p'_0) of isotropically-consolidated undrained compression (CIUC)

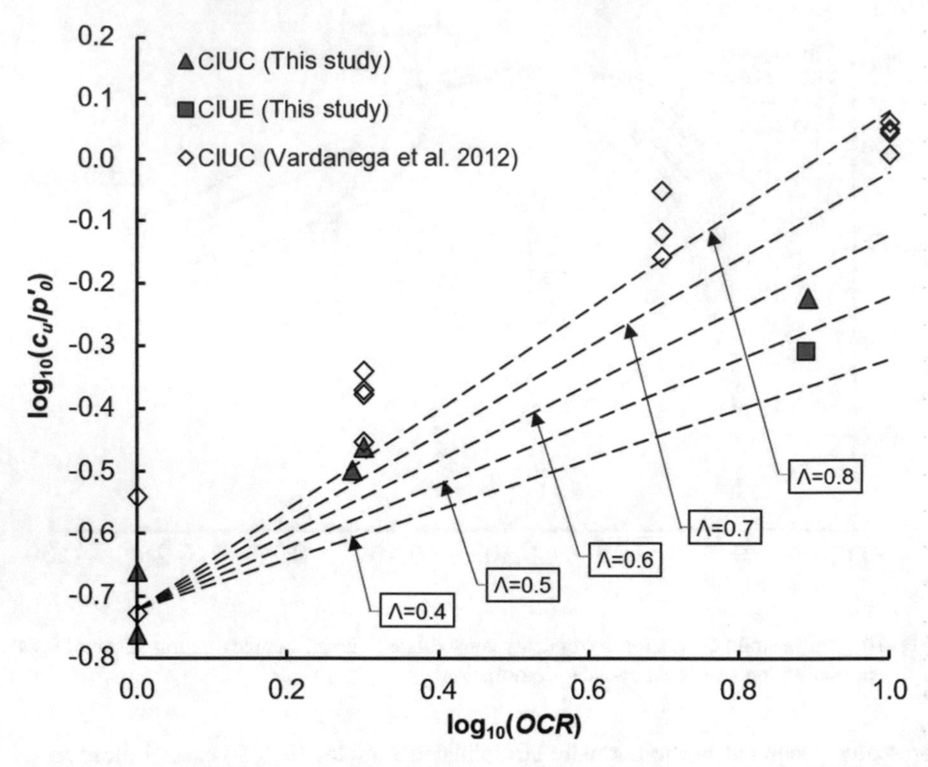

Fig. 11. Relationship between undrained shear strength ratio (cylinder) and *OCR* (following the frameworks of Ladd et al. 1977 and Mayne 1980)

tests appears to increase with *OCR*. Following Eq. 1, the parameter Λ obtained by regression is 0.60 (for $(c_u/p'_0)_{NC} = 0.19$, $n = 6$ and $R^2 = 0.92$), which falls within the range of Λ (0.130–0.998) observed by Mayne (1980) and is slightly lower than the average value (0.70) found for CIUC tests on a variety of soils (Mayne 1988). While the strength ratios of compression tests measured in this study are slightly lower than those found from CIUC tests on similar kaolin material (Vardanega et al. 2012), the values of Λ are similar. Figure 11 also shows the single CIUE test for comparison.

In Fig. 12, the deformation parameter $\gamma_{M=2}$ for each test is presented to examine the effect of overconsolidation on measured shear strain. The results indicate that mobilised shear strain increases with *OCR*, which agrees with the positive trend found by Vardanega et al. (2012) for similar tests on another kaolin material.

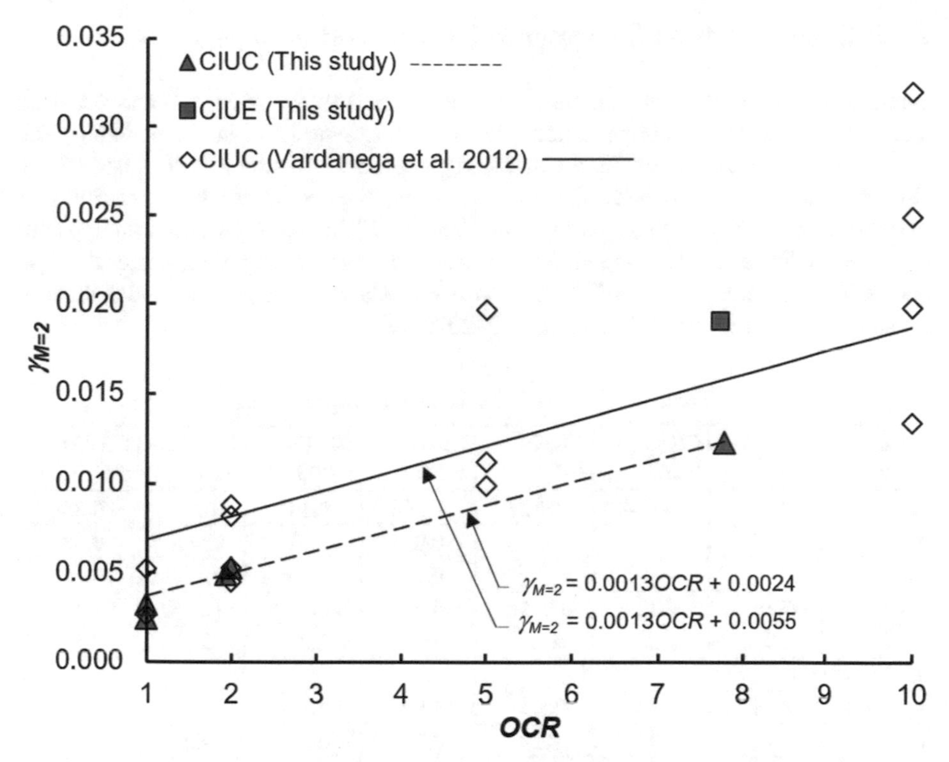

Fig. 12. Dependence of deformation parameter $\gamma_{M=2}$ (cylinder) on *OCR* values Values from Vardanega et al. 2012 for *OCR* = 15 and 20 not shown but best fit linear line generated using the entire dataset. Best fit line through the new CIUC data from this study does not include the single CIUE test.

6 Discussion

6.1 Influence of Test Conditions

Figure 3 and Table 4 suggest that values of λ are sensitive to initial void ratio for isotropic samples. A smaller range in λ is observed for the oedometer tests when compared to isotropic data, given the same range in initial void ratio. The rate of isotropic consolidation appears to have relatively little influence on measured compressibility parameters.

Examination of the stress paths in Fig. 7 and normalised stress-strain curves in Fig. 10 reveal that the large strain behaviour of CIUC-2-b-200 may have been significantly affected by bedding at the start of shear. A 'concave' load cap was used, which caused sudden displacement and changes in measured load when axial alignment was forced in the early stages of undrained shearing. The use of a rotationally flexible load cap, however, appears to have little influence on the mobilised strains (observed in Fig. 10) measured in undrained compression- although some differences in the effective stress behaviour are noticeable in Fig. 7.

7 Influence of Data Representation for Test Analysis

Comparison of the isotropic compression data (triaxial) shown in Figs. 3 and 4 reveals that the bilogarithmic representation yields no apparent increase in linearity. The coefficients of determination for the deduced parameters of λ and λ^* and κ and κ^* are almost identical for this dataset. However, replotting the oedometer data from Fig. 5 in bilogarithmic form (as suggested by Butterfield 1979) in Fig. 6 produces an apparent increase in linearity. The regression equations shown in Fig. 6 indicate that the bilogarithmic representation of K_0 compression data can adequately model the full stress range for each test (10–200 and 10–2000 kPa).

Table 5. Summary of state and shear parameters

Triaxial Test:	CIUC-1-a-395	CIUC-1-b-200	CIUC-1-b-403	CIUC-2-a-208	CIUC-2-b-200	CIUC-8-a-51	CIUE-8-a-52
λ	0.206	0.124	0.127	0.164	0.141	0.195	0.258
κ	–	–	–	0.034	0.019	0.045	0.023
Λ	–	–	–	0.793	0.865	0.769	0.911
p'_0	395.16	200.12	403.15	208.04	200.08	51.3	51.73
OCR	1.0	1.0	1.0	1.9	2.0	7.8	7.7
Right cylinder							
c_u (kPa)	68.25	43.63	75.41	65.90	69.12	30.62	−25.49
c_u/p'_0	0.17	0.22	0.19	0.32	0.35	0.60	0.49
p'_{peak} (kPa)	197.14	111.00	216.18	166.89	155.57	68.76	57.89
M	0.69	0.79	0.70	0.79	0.89	0.89	−0.88
$\varepsilon_{a\ peak}$ (%)	4.97	6.05	5.63	6.16	10.49	7.58	−9.24
$\gamma_{M\ =\ 2}$	0.0032	0.0025	0.0051	0.0049	0.0053	0.0123	0.0146
b	0.335	0.263	0.406	0.341	0.293	0.416	0.413
Parabolic bulging or necking							
c_u (kPa)	66.66	42.48	73.43	64.04	65.71	29.49	−26.69
c_u/p'_0	0.17	0.21	0.18	0.31	0.33	0.57	0.52
p'_{peak} (kPa)	202.29	112.32	219.25	169.26	154.05	67.83	57.09
M	0.66	0.76	0.67	0.76	0.85	0.87	−0.94
$\varepsilon_{a\ peak}$ (%)	4.40	5.32	4.98	5.36	9.83	7.23	−9.24
$\gamma_{M=2}$	0.0030	0.0022	0.0048	0.0046	0.0045	0.0114	0.0161
b	0.339	0.263	0.413	0.344	0.305	0.412	0.412

During undrained shear of a triaxial sample, the deformation shape is commonly assumed to be either cylindrical or parabolic, due to the variable influence of frictional end restraint. Table 5 provides the values of parameters for strength (c_u) and strain ($\gamma_{M=2}$) for each triaxial test deduced using the assumption of cylindrical or parabolic sample shape. In compression tests, the parabolic assumption estimates up 5% reduction in strength compared to employing the assumption of right cylinder; while up to 17% reduction is observed in $\gamma_{M=2}$. The extension strength is calculated to be 5% greater with necking than the value calculated assuming cylindrical deformation; similarly, $\gamma_{M=2}$ increases by 9% with the same assumption. This suggests that the strength and strain parameters used in design calculations may be affected by the assumed shape of the sample during shear.

Figures 3 and 5 demonstrate considerable differences in behaviour between oedometer and isotropic samples of similar material. If different values of λ and κ for each of the isotropic and K_O test series are selected as 'representative' compressibility parameters, the deviation in predicted c_u can be estimated using Eq. 1 in combination with Eq. 4 (which is based on critical state soil mechanics i.e. Schofield and Wroth 1968) of the form shown in Muir Wood (1990):

$$\Lambda = \frac{\lambda_{-\kappa}}{\lambda} \tag{4}$$

Table 6 shows that the chosen 'representative' values of Λ vary slightly between isotropic and K_O test data. These values of Λ, calculated using Eq. 4, are considerably higher than the value (0.60) obtained by regression of measured undrained shear strength ratios (shown in Fig. 11). It is possible that the lower value of Λ obtained using Eq. 1 may be due a lower than expected c_u value for the OCR = 8 compression test.

Table 6. Derived parameters for c_u prediction (right cylinder assumed)

Consolidation and swelling parameters derived from:	Isotropic compression (Average)	(n)	K_O compression	Measured c_u (Eq. 1)
λ	0.160	(6)	0.231	–
κ	0.033	(3)	0.057	–
Λ	0.795		0.753	0.60

8 Conclusions

The following conclusions are observed from the experimental results in this paper:

- The results of compression tests suggest that isotropic consolidation parameters are sensitive to initial void ratio, possibly due to varying degrees of swelling during saturation. Values of λ and κ from oedometer tests on non-preconsolidated material are less sensitive to initial void ratio and demonstrate an increase in linearity when represented in bilogarithmic form.
- From a small number of triaxial samples sheared in compression, it appears that the type of load cap connection can have a significant influence on the large strain behaviour of kaolin samples.

- Up to 5% and 17% variation in observed c_u and $\gamma_{M=2}$ can arise from the assumption of sample deformation shape (cylindrical versus parabolic)
- The variations of strength ratio and shear strain with overconsolidation ratio appear to be described by positive trends, which agree with the results of similar, previously published tests on kaolin.

Acknowledgments. The first author would like to express her appreciation for the help and support she received from the late Dr David Nash. The first author also gratefully acknowledges the DTP scholarship provided by the University of Bristol. The authors thank Mr. Gary Martin for his technical services and support.Notation ListThe following notation is used in this paper

B	Skempton's B-value;
b	an exponent (Eq. 2);
$CIUC$	isotropically consolidated undrained compression test;
$CIUE$	isotropically consolidated undrained extension triaxial test;
c_u	undrained shear strength;
e_m	void ratio measured at the end of consolidation under p'_m;
e_0	void ratio measured at the end of consolidation under p'_0;
Δe_{PRE}	change in void ratio measured while the load was held "pre-shear" i.e. between the end of consolidation (and swelling) and the start of shearing;
Δe_{SAT}	change in void ratio during sample saturation;
G_S	specific gravity;
n	number of data points in a series or regression;
nc	normally consolidated;
OCR	overconsolidation ratio;
oc	overconsolidated
p'	mean effective stress;
p'_m	maximum mean effective stress during consolidation;
p'_0	mean effective stress after swell back;
p'_{sat}	mean effective stress during sample saturation;
q	deviator stress;
R^2	coefficient of determination of a correlation;
RD	relative deviation (as defined in Waters and Vardanega 2009);
v	specific volume;
w_L	liquid limit;
w_P	plastic limit;
γ	shear strain, (1.5 times the axial strain (ε_a) is used in this paper);
$\gamma_{M=2}$	shear strain mobilised at $0.5c_u$;
ε_a	axial strain;
$\varepsilon_{a\ \text{SAT}}$	axial strain measured from extrusion to the end of saturation
κ	slope of swelling line in semi-logarithmic space;
κ^*	slope of swelling line in bilogarithmic space;
Λ	an exponent (Eq. 1);
λ	slope of normal compression line in semi-logarithmic space;

λ^*	slope of normal compression line in bilogarithmic space;
σ'_{vm}	maximum past effective vertical stress in the ground;
σ'_{v0}	vertical effective stress in the ground;
τ_{mob}	mobilised shear stress.

References

Bialowas, G.: Time and stress dependent mechanical properties of reconstituted chalk. Ph.D. thesis, University of Bristol, Bristol, United Kingdom (2016)

Bishop, A.W., Henkel, D.J.: The Measurement of Soil Properties in the Triaxial Test. Edward Arnold, London (1957)

Bjerrum, L.: Problems of soil mechanics and construction of soft clays and structurally unstable soils (collapsible, expansive and others). General report, Session four. In: Proceedings 8th International Conference on Soil Mechanics and Foundation Engineering, Moscow, 6–11 August 1973, vol. 3, pp. 111–159 (1973)

BSI: Methods of test for soils for civil engineering purposes. British standard BS1377. British Standards Institution, London, United Kingdom (1990)

Butterfield, R.: A natural compression law for soils (an advance on e–log p′). Géotechnique **29** (4), 469–480 (1979). https://doi.org/10.1680/geot.1979.29.4.469

Cerato, A.B., Lutenegger, A.J.: Determining intrinsic compressibility of fine-grained soils. J. Geotech. Geoenviron. Eng. **130**(8), 872–877 (2004). https://doi.org/10.1061/(ASCE)1090-0241(2004)130:8(872)

Jamiolkowski, M., Ladd, C.C., Germaine, J.T., Lancellotta, R.: New developments in field and laboratory testing of soils. state of the art. In: Proceedings of the 11th International Conference on Soil Mechanics and Foundation Engineering, San Francisco, 12–16 August 1985, pp. 57–153 (1985)

Ladd, C.C., Foott, R.: A new design procedure for stability of soft clays. J. Geotech. Eng. Div. (Am. Soc. Civil Eng.) **100**(7), 763–786 (1974)

Ladd, C., Foot, R., Ishihara, K., Schlosser, F., Poulos, H.: Stress-deformation and strength characteristics. In: Proceedings of the 9th International Conference on Soil Mechanics and Foundation Engineering, Tokyo, vol. 2, pp. 421–494 (1977)

Lade, P.V.: Triaxial Testing of Soils. John Wiley & Sons Ltd., Chichester (2016)

Mayne, P.W.: Cam-Clay predictions of undrained strength. J. Geotech. Eng. Div. (Am. Soc. Civil Eng.) **106**(11), 1219–1242 (1980)

Mayne, P.W.: Determining OCR in clays from laboratory strength. J. Geotech. Eng. (Am. Soc. Civil Eng.) **114**(1), 76–92 (1988). https://doi.org/10.1061/(ASCE)0733-9410(1988)114:1(76)

Muir Wood, D.: Soil Behaviour and Critical State Soil Mechanics. Cambridge University Press, Cambridge (1990)

Sarsby, R.W., Kalteziotis, N., Haddad, E.H.: Bedding error in triaxial tests on granular media. Géotechnique **30**(3), 302–309 (1980). https://doi.org/10.1680/geot.1980.30.3.302

Schofield, A.N., Wroth, C.P.: Critical State Soil Mechanics. McGraw-Hill Maidenhead, United Kingdom (1968)

Skempton, A.W.: The pore-pressure coefficients A and B. Géotechnique **4**(4), 143–147 (1954). https://doi.org/10.1680/geot.1954.4.4.143

Sukolrat, J.: Structure and destructuration of Bothkennar clay. Ph.D. thesis, University of Bristol, Bristol, United Kingdom (2007)

Vardanega, P.J., Bolton, M.D.: Strength mobilization in clays and silts. Can. Geotech. J. **48**(10), 1485–1503 (2011). https://doi.org/10.1139/t11-052 and Corrigendum, 49(5), 631, https://doi.org/10.1139/t2012-023

Vardanega, P.J., Lau, B.H., Lam, S.Y., Haigh, S.K., Madabhushi, S.P.G., Bolton, M.D.: Laboratory measurement of strength mobilisation in kaolin: link to stress history. Géotechnique Lett. **2**(1), 9–15 (2012). https://doi.org/10.1680/geolett.12.00003

Waters, T.J., Vardanega, P.J.: Re-examination of the coefficient of determination (r2) using road materials engineering case studies. Road Transp. Res. **18**(3), 3–12 (2009)

Wroth, C.P., Loudon, P.A.: The correlation of strains within a family of triaxial tests on overconsolidated samples of kaolin. In: Proceedings Geotechnical Conference, Oslo, pp. 159–163 (1967)

Explicit Model for Excess Porewater Pressure Computation in Fine-Grained Soils with Arbitrary Initial Conditions

Amir Al-Khafaji(✉)

Department of Civil Engineering and Construction,
Bradley University, Peoria, IL, USA
amir@fsmail.bradley.edu

Abstract. The consolidation process in fine-grained soils involves changes in excess pore water pressure (EPWP) and void ratio with time. An analytical solution for this problem was presented by Terzaghi (1943). Their solutions included several simplifying assumptions and applied to a constant initial EPWP distributions with depth. The solution is widely used for educational purposes and is of limited practical use where the initial EPWP is complex. The Terzaghi solution is not valid for complex and/or arbitrary applied loading. In such cases, numerical methods are used for EPWP calculation versus time. The finite differences and the finite element methods are powerful tools that are effectively used for modeling the compression behavior of fine-grained soil deposits. Unfortunately, these methods often require large number of calculations and suffer from round-off errors when large number of time increments is required. The proposed model minimizes the round-off errors and eliminates the large number of calculations associated with numerical techniques. The basis of the proposed method is the development of eigenvalues and eigenvectors associated with time-rate of settlement. The proposed explicit model in effect produces a direct solution at any given time and depth. The solution permits the evaluation of EPWP at even a fraction of a time increment. Two models are presented for doubly-drained soil layers and singly-drained layers. Unlike the finite differences method, the proposed model allows arbitrary initial EPWP distribution and for any number of time increments by multiplying four matrices involving soil properties, time increment used and the initial EPWP.

Keywords: Consolidation · Finite differences · Round-off · Void ratio
Load · Soils · Increment · EPWP · Analytical · Numerical

1 Introduction

Although analytical methods will continue to provide useful solution tro then time-rate of settlement problem, they cannot yield realistic answers for problems involving arbitrary initial conditions. Most time rate of settlement estimates are based on a one-dimensional model using an initial applied vertical stress distribution. Consolidation tests are used for the purpose of determining soil compression properties. The primary consolidation process involves changes in excess pore water pressure (EPWP) and

J. S. McCartney and L. R. Hoyos (Eds.): GeoMEast 2018, SUCI, pp. 119–131, 2019.
https://doi.org/10.1007/978-3-030-01914-3_10

subsequently in void ratio with time. The one-dimensional model which relates EPWP distributions u at depth, z and at any time, t is given by

$$\frac{\partial u}{\partial t} = c_v \frac{\partial^2 u}{\partial z^2} \tag{1}$$

A closed form solution for this problem has been presented by Terzaghi and Frolich (1936). Their solutions included several simplifying assumptions and applied to a constant initial EPWP distribution with depth. For most practical problems, the initial EPWP distribution is non-linear with depth due to partial EPWP dissipation under previous load and/or surface load such as spread footings that produce Boussinesq initial excess porewater pressure. Consequently, numerical methods are generally used to solve Eq. 1.

2 The Finite Difference Method For Doubly-Drained Layer

The basic concept involves discretization of arbitrary continuous functions and replacing them with polynomials whose derivatives are used in approximating mathematical model behaviors. The solution of Eq. (1) at a given time $t = t_j$ and depth z_i. The first derivative of the arbitrary function $u(z, t_j)$ at node z_i can be approximated using a backward node, a forward node, or a backward and a forward node. These are referred to as backward, forward, and central derivatives respectively. The implication in all cases is that a first order polynomial is used to approximate the first derivative of any arbitrary function $u(z, t_j)$. The backward, forward, and central derivatives at $t = t_j$ and $z = z_i$ are given respectively as follows;

$$\left.\frac{\partial u}{\partial z}\right|_{@t_i \& z_i} = \frac{-u_{i-1,j} + u_{i,j}}{\Delta z} \tag{2a}$$

$$\left.\frac{\partial u}{\partial z}\right|_{@t_i \& z_i} = \frac{-u_{i,j} + u_{i+1,j}}{\Delta z} \tag{2b}$$

$$\left.\frac{\partial u}{\partial z}\right|_{@t_i \& z_i} = \frac{-u_{i-1,j} + u_{i+1,j}}{2\Delta z} \tag{2c}$$

It can be shown that the central derivative is more accurate than the backward and the forward derivative approximations (Al-Khafaji and Tooley 1986). The second central derivative is expressed in terms of the rate of change of the backward and forward derivatives. Hence, at node z_i and at $t = t_j$, write

$$\left.\frac{\partial^2 u}{\partial z^2}\right|_{@t_j \& z_i} = \frac{-(\text{backward Derivative})_j + (\text{Forward Derivative})_j}{\Delta z}$$

$$\left.\frac{\partial^2 u}{\partial z^2}\right|_{@t_j\&z_i} = \frac{1}{\Delta z}\left[-\left(\frac{-u_{i-1,j}+u_{i,j}}{\Delta z}\right)+\left(\frac{-u_{i,j}+u_{i+1,j}}{\Delta z}\right)\right] = \frac{u_{i-1,j}-2u_{i,j}+u_{i+1,j}}{(\Delta z)^2} \quad (3)$$

The approximating of the first derivative of u with respect to time is approximated at $t = t_j$ and $z = z_i$ in using the forward derivative approximation. The implication is that given initial EPWP distribution with depth, a new EPWP values at a new time increment can be calculated. Thus, the first order partial derivative term appearing in Eq. 1 can be expressed in difference form as

$$\left.\frac{\partial u}{\partial z}\right|_{@t_j\&z_i} = \frac{-u_{i,j} + u_{i,j+1}}{\Delta t} \quad (4)$$

Substituting these expressions into equations into Eq. 1 gives

$$\frac{-u_{i,j} + u_{i,j+1}}{\Delta t} = c_v \frac{u_{i-1,j} - 2u_{i,j} + u_{i+1,j}}{(\Delta z)^2} \quad (5a)$$

For a soil layer of thickness Ho, the depth increment $\Delta z = Ho/m$ and the time increment $\Delta t = t/n$. Where m is the number of depth increments and n is the number of time increments for which EPWP is required. In general, $i = 0, 1, \ldots, m$ and $j = 0, 1, \ldots, n$. Denoting $\alpha = c_v\Delta t/\Delta z^2$, then simplifying and rearranging Eq. (5a) yields;

$$u_{i,j+1} = \alpha u_{i-1,j} - (1-2\alpha)u_{i,j} + \alpha u_{i+1,j} \quad (5b)$$

Equation (5b) is an explicit finite difference recurrence formula which permits direct step-by-step evaluation of EPWP at specified number of time increment. Thus, knowing the initial and boundary EPWP values at $t = 0$, it is possible to calculate the EPWP at new time increments $t = \Delta t, 2\Delta t, \ldots, n\Delta t$. We now consider the case of a doubly-drained layer as shown in Fig. 1.

Using Eq. (5b), one may write a set of linear algebraic equations which can be expressed in a matrix form for the m + 1 nodes along the depth axis z as follows:

$$\left\{\begin{matrix} u_{top} \\ u_1 \\ u_2 \\ \vdots \\ u_{m-1} \\ u_{bot} \end{matrix}\right\}_{t=t_{j+1}} = \begin{bmatrix} \alpha & 1-2\alpha & \alpha & & & & \\ & \alpha & 1-2\alpha & \alpha & & \mathbf{0} & \\ & & \ddots & \ddots & \ddots & & \\ & & & \alpha & 1-2\alpha & \alpha & \\ & \mathbf{0} & & & \alpha & 1-2\alpha & \alpha \end{bmatrix} \left\{\begin{matrix} u_{top} \\ u_1 \\ u_2 \\ \vdots \\ u_{m-1} \\ u_{bot} \end{matrix}\right\}_{t=t_j} \quad (6)$$

For a soil layer draining at its top and bottom boundaries, the EPWP drops to zero after the first time increment. For the special case shown in Fig. 1, at $t = t_1$ we have $u_{top} = u_{bot} = 0$ and Eq. (6) is simplified to the following:

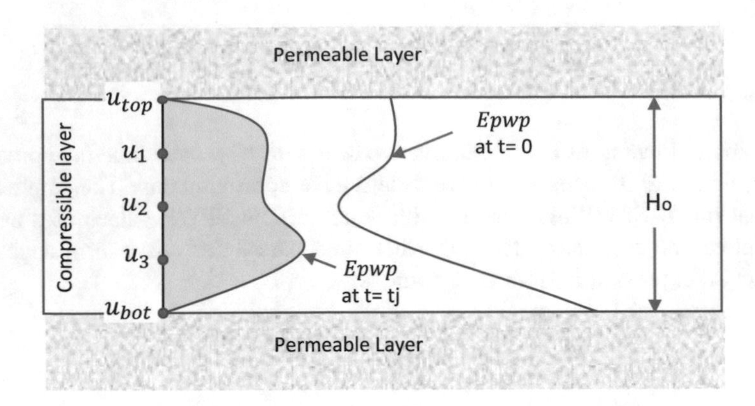

Fig. 1. Schematic representation of a soil layer with double drainage.

$$\begin{Bmatrix} u_1 \\ u_2 \\ u_3 \end{Bmatrix}_{t=t_2} = \begin{bmatrix} 1-2\alpha & \alpha & 0 \\ \alpha & 1-2\alpha & \alpha \\ 0 & \alpha & 1-2\alpha \end{bmatrix} \begin{Bmatrix} u_1 \\ u_2 \\ u_3 \end{Bmatrix}_{t=t_1} \tag{7}$$

Equation (7) can be written more conveniently in a compact matrix form as

$$\{u\}_{j+1} = [A]\{u\}_j \qquad j = 1, 2, \ldots, n \tag{8}$$

The value of α may not exceed 0.50 for the solution to be stable, in fact $\alpha = 1/6$ gives the most accurate results (Scott 1963). Note that Eq. (8) applies after the first time increment because of discrepancies between the initial and boundary conditions. It is evident that the limitation imposed on α makes it necessary that the EPWP be evaluated at large number of time increments. Thus, round-off errors, truncation errors and computational effort could be problematic.

3 The Finite Difference Method For Singly-Drained Layer

For a singly-drained soil layer Eq. (5b) is still valid and permits direct step-by-step evaluation of EPWP. The implication is that knowing the initial and boundary EPWP values at $t = 0$, it is possible to calculate the EPWP at new time increments $t = \Delta t, 2\Delta t, \ldots, n\Delta t$. Consider the singly-drained soil layer shown in Fig. 2.

Using Eq. (5b), one may write a set of linear algebraic equations which can be expressed for the m + 1 nodes along the depth axis z. However, the EPWP at the bottom of the layer is unknown. That is, $u_{top} = 0$ and u_{bot} is unknown. Thus, at each node at time $t = \Delta t = t_1$ yield:

Node 1: $u_{1,1} = \alpha u_{top,0} + (1 - 2\alpha)u_{1,0} + \alpha u_{2,0}$
Node 2: $u_{2,1} = \alpha u_{1,0} + (1 - 2\alpha)u_{2,0} + \alpha u_{3,0}$
Node 3: $u_{3,1} = \alpha u_{2,0} + (1 - 2\alpha)u_{3,0} + \alpha u_{bot,0}$
Node bottom: $u_{bot,1} = \alpha u_{3,0} + (1 - 2\alpha)u_{bot,0} + \alpha u_{5,0}$

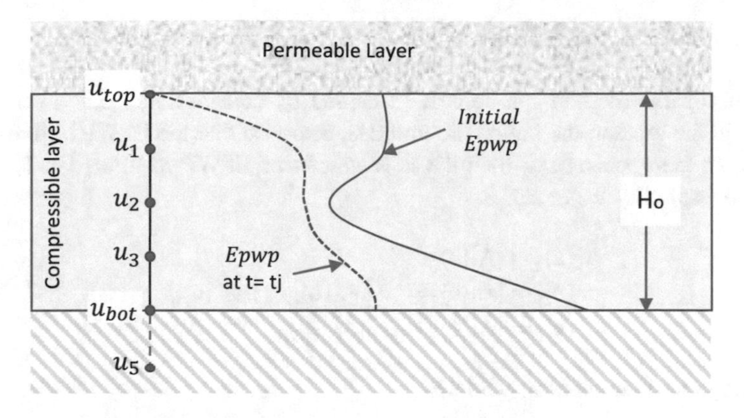

Fig. 2. Schematic representation of a soil layer with single drainage.

Note that, we must now remove the fictitious node u_5 using the boundary condition that the seepage at the bottom is equal to zero. That is

$$k\left(\frac{\partial u}{\partial z}\right)_{t=o} = k\frac{-u_{3,0} + u_{5,0}}{2\Delta z} = 0$$

This gives

$$u_{3,0} = u_{5,0}$$

Therefore, the fictitious EPWP at node 5 is replaced by the real EPWP at node 3. Thus, $u_{bot,1} = 2\alpha u_{3,0} + (1 - 2\alpha)u_{bot,0}$. In a matrix form and at $t = t_1$, $u_o = u_{top} = 0$

$$\begin{Bmatrix} u_1 \\ u_2 \\ u_3 \\ u_{bot} \end{Bmatrix}_{t=t_1} = \begin{bmatrix} 1-2\alpha & \alpha & 0 & 0 \\ \alpha & 1-2\alpha & \alpha & 0 \\ 0 & \alpha & 1-2\alpha & \alpha \\ 0 & 0 & 2\alpha & 1-2\alpha \end{bmatrix} \begin{Bmatrix} u_1 \\ u_2 \\ u_3 \\ u_{bot} \end{Bmatrix}_{t=0} \tag{9}$$

Equation (9) can be written more conveniently in a compact matrix form as

$$\{u\}_{j+1} = [B]\{u\}_j \qquad \text{for } j = 1, 2, \ldots, n \tag{10}$$

Note that the matrix [B] shown in Eq. (10) is different form matrix [A] given by Eq. (8). The value of α may not exceed 0.50 for the solution to be stable. Also, because of discrepancies between the initial and boundary conditions the solution begins with the EPWP being evaluated after first time increment.

4 The Proposed Method

The basis for this method can be best explained by considering a soil layer with free drainage at the top and the bottom boundaries. Suppose that the EPWP initially at t = 0 is given by the vector $\{u\}_o$, then the new values of EPWP at times j = 1, …, n are computed explicitly using Eq. 8.

$$\begin{aligned} \langle u\rangle_1 &= [A]\langle u\rangle_0 \\ \langle u\rangle_2 &= [A]\langle u\rangle_1 = [A][A]\langle u\rangle_0 = [A]^2\langle u\rangle_0 \\ \langle u\rangle_3 &= [A]\langle u\rangle_2 = [A][A]^2\langle u\rangle_0 = [A]^3\langle u\rangle_0 \end{aligned} \tag{11a}$$

for the anytime increment n, we have

$$\{u\}_n = [A]^{n-j}\{u\}_j \qquad for \quad j = 1,\ldots,n \tag{11b}$$

Equation (11b) has the advantage over Eq. (8) in that it doesn't require the computation of intermediate EPWP values! Raising a square matrix to any power is an eigenproblem. The eigenvalues and their corresponding vectors for matrix [A] can be determined by solving the following equation:

$$[A]\{\phi\} = \lambda\{\phi\} \tag{12}$$

Where λ is the eigenvalue for which an eigenvector is calculated from the square matrix [A] whose size is $(m-1) \times (m-1)$. This is because there are m nodes with depth and the EPWP at the top and bottom are eliminated due to free drainage. Hence, the solution for the $\{\phi\}$ vector is achieved by forcing the following determinant to zero to yield the $m-1$ eigenvalues. Thus

$$|[A] - \lambda[I]| = 0 \tag{13}$$

Expanding the determinant in Eq. (13) yields a polynomial of order $(m-1)$ and whose roots are the eigenvalues. The corresponding eigenvectors $\{\phi\}_1$, $\{\phi\}_2$, …, $\{\phi\}_{m-1}$ are computed using traditional techniques. Note that because free drainage is assumed at the top and bottom of the soil layer, there are $m-1$ depths (nodes) at which the EPWP needs to be computed. Therefore,

$$\begin{aligned} [A]\{\phi\}_1 &= \lambda_1\{\phi\}_1 \\ [A]\{\phi\}_2 &= \lambda_2\{\phi\}_2 \\ &\vdots \\ [A]\{\phi\}_{m-1} &= \lambda_{m-1}\{\phi\}_{m-1} \end{aligned}$$

These equations can be expressed more conveniently in the following form;

$$[A]\left[\{\phi\}_1\{\phi\}_2\cdots\{\phi\}_{m-1}\right] = \left[\{\phi\}_1\{\phi\}_2\cdots\{\phi\}_{m-1}\right]\begin{bmatrix} \lambda_1 & & & \mathbf{0} \\ & \lambda_2 & & \\ & & \ddots & \\ \mathbf{0} & & & \lambda_{m-1} \end{bmatrix} \tag{14a}$$

in a compact matrix form, we have

$$[A][] = [\Phi][\lambda] \tag{14b}$$

where $[\Phi]$ is a square eigenvectors matrix and $[\lambda]$ is a diagonal eigenvalues matrix. Therefore, multiplying Eq. (14b) by the inverse of the eigenvector matrix $[\Phi]^{-1}$ matrix gives;

$$[A] = [\Phi][\lambda][\Phi]^{-1}$$

The square of the matrix [A] is now given as;

$$[A]^2 = [\Phi][\lambda][\Phi]^{-1}[\Phi][\lambda][\Phi]^{-1}$$
$$[A]^2 = [\Phi][\lambda]^2[\Phi]^{-1}$$

Similarly raising [A] to the n-th power yields;

$$[A]^n = [\Phi][\lambda]^n[\Phi]^{-1} \tag{15}$$

Substituting Eq. (15) into Eq. (11b) gives the explicit eigenproblem solution to the one-dimensional time rate of settlement problem with free drainage.

$$\{u\}_n = [A]^{n-j}\{u\}_j = [\Phi_A][\lambda_A]^{n-j}[\Phi_A]^{-1}\{u\}_j \quad \text{for } j = 1, \ldots, n \tag{16}$$

Note that Eq. (16) involves raising a diagonal matrix to the power n which is easily achieved by raising the diagonal elements to that same power. Thus, for a doubly drained layer, the eigenvector matrix is designated by $[\Phi_A]$ and the corresponding vector by $[\lambda_A]$. The advantage of using Eq. (16) over the conventional finite difference procedure is that all intermediate EPWP vectors are eliminated and the solution is achieved by raising a diagonal matrix to any time increment including a fraction of a time increment.

Similarly, the solution for a singly-drained layer can be achieved using the same procedure except the square matrix is defined by a new matrix [B] given by Eq. (10). That is, for singly-drained soil layer, we have

$$\{u\}_n = [B]^{n-j}\{u\}_j = [\Phi_B][\lambda_B]^{n-j}[\Phi_B]^{-1}\{u\}_j \quad \text{for } j = 1, \ldots, n \tag{17}$$

Equation (17) applies for a singly drained layer, the eigenvector matrix is designated by $[\Phi_B]$ and the corresponding eigenvalue matrix by $[\lambda_B]$ to distinguish it from the doubly drained matrix [A] given by Eq. (8).

5 Accuracy For Doubly-Drained Layers

For the proposed technique to be useful, we need to show that its accuracy in computing the EPWP, the Time Factor T and the Average Degree of Consolidation U. For a soil drained at both ends with thickness of Ho, the drainage path Hdp is given by

$$H_{dp} = \frac{H_o}{2} = \frac{m}{2}\Delta z \tag{18a}$$

Where m is the number of depth increments use. The coefficient of consolidation can be expressed next in terms of α and the depth and time increments as follows:

$$c_v \frac{(\Delta z)^2}{\Delta t}\alpha \tag{18b}$$

It is often convenient to express time in terms of the dimensionless parameter known as the time factor T. This is expressed in terms of the length of drainage path H_{dp}, the time in question t, and the coefficient of consolidation c_v as follows:

$$T \frac{c_v t}{H_{dp}^2} \tag{19}$$

Substituting Eq. (18a) and (18b) into Eq. 19 and noting that $t = n\Delta t$ gives the time factor in terms of α, the depth increment Δz, and time increments Δt. Thus

$$T = \frac{\left[\alpha(\Delta z)^2/\Delta t\right](n\Delta t)}{(m\Delta z/2)^2} = 4\frac{\alpha n}{m^2} \tag{20}$$

The average degree of consolidation for the entire soil layer at any time U_j is determined in terms of the initial area enclosed by the EPWP versus depth distribution A_o and the area enclosed by the EPWP versus depth at any time A_j. Thus,

$$U_j = 1 - \frac{A_j}{A_o} \qquad j = 0, \ldots, n \tag{21}$$

Equation (21) yields average degree of consolidation in terms the areas A_o and A_j representing to the areas under the EPWP at t = to and t = tj, respectively. This is achieved using the trapezoidal rule or Simpson's 1/3 Rule of integration.

6 Accuracy For Singly-Drained Layers

For a soil layer drained at one end, the Time Factor T and the Average Degree of Consolidation U can be computed as follows:

$$H_{dp} = H_o = m\Delta z \tag{22}$$

The coefficient of consolidation was given by Eq. (18b). Substituting Eqs. (18b) and (22) into Eq. (20) and noting that t = nΔt gives

$$T = \frac{\frac{(\Delta z)^2}{\Delta t}\alpha(n\Delta t)}{(m\Delta z)^2} = \frac{\alpha n}{m^2} \tag{23}$$

The average degree of consolidation for the entire soil layer at any time U_j is determined using Eq. 21.

Example 1 Determine the average degree of consolidation, time factor, and the EPWP after 7 years for the doubly-drained clay layer shown in Fig. 1. Assume a thickness of 16 m, $c_v = 8.2\ m^2/yr$, and a constant initial EPWP distribution of 60 kPa.

Solution:

Substituting α = 1/6 into Eq. (7) with Δz = 16/4 = 4 m and substituting into Eq. (18b), then solving for the time increment gives

$$\Delta t = \alpha\frac{(\Delta z)^2}{c_v} = \frac{1}{6}\frac{(4)^2}{(8.2)} = 0.3252\ \ year$$

$$\text{n} = 7/0.3252 = 21.525 \text{ increments}$$

It is important to note that the initial EPWP at time t = 0 at the top and bottom boundaries is not equal to zero. However, the EPWP at these boundaries drops to zero at $t = 0^+$. Furthermore, at t = 0, the EPWP $u_{top} = u_{bot} = 60$ kPa. This discrepancy between the boundary conditions and the initial EPWP distribution is resolved by taking the average value of EPWP of 30 kPa as the initial EPWP at the boundaries to start the solution. Thus, we need to solve for the EPWP vector at t = Δt as

$$u_{1,1} = \alpha u_{top,o} + (1 - 2\alpha)u_{1,o} + \alpha u_{2,o} = \tfrac{30}{6} + \left(\tfrac{2}{3}\right)60 + \tfrac{60}{6} = 55\,kpa$$

$$u_{2,1} = \alpha u_{1,o} + (1 - 2\alpha)u_{2,o} + \alpha u_{3,o} = \tfrac{60}{6} + \left(\tfrac{2}{3}\right)60 + \tfrac{60}{6} = 60\,kpa$$

$$u_{3,1} = \alpha u_{2,o} + (1 - 2\alpha)u_{3,o} + \alpha u_{bot,o} = \tfrac{60}{6} + \left(\tfrac{2}{3}\right)60 + \tfrac{30}{6} = 55\,kpa$$

Using MATLAB, the $\{u\}_{21.525}$ vector in Eq. (16) can be calculated in terms of the eigenvectors and eigenvalues using $\alpha = 1/6$. Thus, after 7 years by substituting j = 1 and n = 21.525 gives

$$\begin{Bmatrix} u_1 \\ u_2 \\ u_3 \end{Bmatrix}_{21.525} = \frac{1}{4}\begin{bmatrix} -1 & 1 & 1 \\ 0 & -\sqrt{2} & \sqrt{2} \\ 1 & 1 & 1 \end{bmatrix}\begin{bmatrix} \frac{2}{3} & 0 & 0 \\ 0 & \frac{2}{3}-\frac{\sqrt{2}}{6} & 0 \\ 0 & 0 & \frac{2}{3}+\frac{\sqrt{2}}{6} \end{bmatrix}^{20.525}\begin{bmatrix} -2 & 0 & 2 \\ 1 & -\sqrt{2} & 1 \\ 1 & \sqrt{2} & 1 \end{bmatrix}\begin{Bmatrix} 55 \\ 60 \\ 55 \end{Bmatrix}_1$$
$$= \begin{Bmatrix} 6.22 \\ 8.79 \\ 6.22 \end{Bmatrix}_1$$

Note that EPWP is zero at the top and bottom of the clay layer and the calculated EPWP vector is given as $\{u\}^t_{21.525} = \{0, 6.22, 8.79, 6.22, 0\}$. The time factor is calculated by substituting m = 4, $\alpha = 1/6$, and n = 21.525 into Eq. (18b) which yields T = 0.8969. The average degree of consolidation is computed next using Eq. (21) and Simpson's 1/3 rule of integration. That is

$$U = 1 - \frac{(\Delta z/3)\left(u_{top} + 4u_1 + 2u_2 + 4u_3 + u_{1bot}\right)_n}{(\Delta z/3)\left(u_{top} + 4u_1 + 2u_2 + 4u_3 + u_{1bot}\right)_0}$$

$$U = 1 - \frac{[0 + 4(6.22) + 2(8.79) + 4(6.22) + 0]_{21.525}}{(60 * 16)_0} = 90.65\%$$

The time factor and the average degree of consolidation compare favorably with the analytical values of T = 0.90 and U = 91.20% reported by Perloff and Baron 1976.

Example 2 Determine the average degree of consolidation, time factor, and the EPWP after 7 years for the singly-drained clay layer shown in Fig. 2. Assume a thickness of 16 m, $c_v = 8.2\ m^2/yr$, and a constant initial EPWP distribution of 60 kPa.

Solution:

Substituting $\alpha = 1/6$ into equation with $\Delta z = 16/4 = 4$ m into Eq. (18b), then solving for the time increment gives

$$\Delta t = \alpha\frac{(\Delta z)^2}{c_v} = \frac{1}{6}\frac{(4)^2}{(8.2)} = 0.3252\ year$$
$$\text{n} = 7/0.3252 = 21.525 \text{ increments}$$

Note that the initial EPWP at time t = 0 at the top boundary is equal to 60 and it drops to zero at $t = 0^+$. This discrepancy between the boundary conditions and the initial EPWP distribution at the top of the layer is resolved by taking the average value of EPWP of 30 kPa as the initial EPWP at the boundaries to start the solution. Thus, the EPWP vector at t = Δt, namely$\{u\}_1$ is determined using α = 1/6 as follows:

$$u_{1,1} = \alpha u_{top,o} + (1-2\alpha)u_{1,o} + \alpha u_{2,o} = \tfrac{30}{6} + \left(\tfrac{2}{3}\right)60 + \tfrac{60}{6} = 55\,kpa$$
$$u_{2,1} = \alpha u_{1,o} + (1-2\alpha)u_{2,o} + \alpha u_{3,o} = \tfrac{60}{6} + \left(\tfrac{2}{3}\right)60 + \tfrac{60}{6} = 60\,kpa$$
$$u_{3,1} = \alpha u_{2,o} + (1-2\alpha)u_{3,o} + \alpha u_{bot,o} = \tfrac{60}{6} + \left(\tfrac{2}{3}\right)60 + \tfrac{60}{6} = 60\,kpa$$
$$u_{bot,1} = 2\alpha u_{2,o} + (1-2\alpha)u_{3,o} = 2\tfrac{60}{6} + \left(\tfrac{2}{3}\right)60 = 60\,kpa$$

Using MATLAB, the EPWP in Eq. (17) can be calculated in terms of the eigenvectors and eigenvalues of matrix [B] using α=1/6.

$$[B] = \begin{bmatrix} 2/3 & 1/6 & 0 & 0 \\ 1/6 & 2/3 & 1/6 & 0 \\ 0 & 1/6 & 2/3 & 1/6 \\ 0 & 0 & 1/3 & 2/3 \end{bmatrix}$$

$$[\lambda_B] = \begin{bmatrix} 0.9746 & 0 & 0 & 0 \\ 0 & 0.7942 & 0 & 0 \\ 0 & 0 & 0.5391 & 0 \\ 0 & 0 & 0 & 0.3587 \end{bmatrix}$$

$$[\Phi_B] = \begin{bmatrix} 0.2420 & -0.5843 & 0.5843 & -0.2420 \\ 0.4472 & -0.4472 & -0.4472 & 0.4472 \\ 0.5843 & 0.2420 & -0.2420 & -0.5843 \\ 0.6325 & 0.6325 & 0.6325 & 0.6325 \end{bmatrix}$$

$$[\Phi_B]^{-1} = \begin{bmatrix} 0.3025 & 0.5590 & 0.7304 & 0.3953 \\ -0.7304 & -0.5590 & 0.3025 & 0.3953 \\ 0.7304 & -0.5590 & -0.3025 & 0.3953 \\ -0.3025 & 0.5590 & -0.7304 & 0.3953 \end{bmatrix}$$

Thus, after 7 years by substituting j = 1 and n = 21.525 into Eq. (17) gives

$$\begin{Bmatrix} u_1 \\ u_2 \\ u_3 \\ u_{bot} \end{Bmatrix}_{21.525} = [\Phi_B][\lambda_B]^{20.525}[\Phi_B]^{-1} \begin{Bmatrix} 55 \\ 60 \\ 60 \\ 60 \end{Bmatrix}_1 = \begin{Bmatrix} 16.96 \\ 31.17 \\ 40.50 \\ 43.73 \end{Bmatrix}$$

The time factor is calculated by substituting m = 4, α = 1/6 and n = 21.525 into Eq. (23) which yields,

$$T = \frac{\alpha n}{m^2} = \frac{1}{6}\left(\frac{21.525}{4^2}\right) = 0.2242$$

The average degree of consolidation is computed next using Eq. (21) and Simpson's 1/3 rule of integration. That is

$$U = 1 - \frac{[0 + 4(16.96) + 2(31.17) + 4(40.50) + 43.73]_{21.525}}{(60 * 16)_0} = 53.34\%$$

The time factor and the average degree of consolidation compare favorably with the analytical values of T = 0.221 and U = 53% reported by Perloff and Baron (1976).

7 Conclusions

The one-dimensional consolidation time rate of settlement problem often involves nonlinear initial EPWP distributions that can be solved numerically. Traditional numerical methods such as finite elements and finite differences techniques require significant number of calculations and suffer from round-off errors. A new technique is presented that is both efficient and versatile and applies to any initial EPWP and boundary conditions. Unlike the traditional numerical techniques, once a model is developed, then arbitrary initial EPWP distribution can be calculated without the need for reworking the problem. Using the finite difference, one would be required to solve intermediate EPWP vectors and repeat the solution for a new initial EPWP. This is not the case with the proposed models. Two models are presented for singly- drained and doubly-drained soil layers with arbitrary initial EPWP distributions. The results compared favorably with exact values for known initial EPWP distributions. Furthermore, the models presented in this paper permits the determination of EPWP at any time including a fraction of a time increment without the need for computing intermediate EPWP values. This eliminate the need for interpolating when EPWP values between increments are required. This is the case when dealing with the finite difference and finite elements methods. The proposed method reduces substantially the roundoff error and computational effort associated with other numerical techniques. The solution at any time is given in terms of an initial EPWP vector and is achieved by multiplying four matrices.

Better models can be developed using greater number of depth increments to more closely describe the complex initial EPWP distributions. Such models will permit more accurate results for the EPWP at any time t. The methods presented can be extended to 2-dimensional and 3-dimensional models. The eigenproblem technique can also be to extended to multi-layer problems involving soils of different compression properties. The proposed method eliminates the need for computation of inconsequential intermediate EPWP values required when using the finite difference method and reduce round-off errors. Geotechnical engineers are no longer encumbered by the procedural details associated with numerical techniques which obscures the broader and intellectually and practically important aspects of time-rate of settlement calculations.

Acknowledgments. Dedicated to my late professor Orlando B. Andersland who taught me the joy and value of scholarship.

References

Al-Khafaji, A.W.N., Tooley, J.R.: Numerical Methods in Engineering Practice. Holt, Rinehart and Winston Inc., New York (1986)

Perloff, W.H., Baron, W.: Soil Mechanics. The Ronald Press Company, New York (1976)

Scott, R.F.: Principles of Soil Mechanics. Addison Wesley Publishing Company Inc, Reading, Massachusetts (1963)

Terzaghi, K.: Theoretical Soil Mechanics. Wiley, New York (1943)

The Effect of Environmental Factors on the Stability of Residential Buildings Built on Expansive Clays

Hossein Assadollahi[1,2(✉)] and Hossein Nowamooz[1]

[1] Department of Civil Engineering and Energies, ICUBE, UMR7357, CNRS, INSA, 67000 Strasbourg, France
hossein.assadollahi@insa-strasbourg.fr
[2] Department of Engineering and Consulting, DETERMINANT R&D SARL, 75008 Paris, France

Abstract. The climate change and the presence of common environmental factors such as extreme drought events, vegetation and root water uptake by trees are considered as serious factors affecting the stability of civil and geotechnical engineering constructions. The present study, investigates the influence of these factors on the structural behavior of residential buildings built on unsaturated expansive clays in field scale. A damaged residential building constructed on shrink-swell clays exposed to root water uptake by trees and also exposed to different climatic conditions over time, has been monitored since 2011 in the south west of France. The crack openings on different sides of the building were monitored using displacement sensors. The soil water content and suction were also monitored in two different angles (north and south) of the building which were close to the trees and the building. The results of the monitoring show that the root water uptake influence the north angle of the building however, the south angle is mostly dominant by the climatic conditions. The difference between the cracks opening movements in each angle of the building is the results of the difference between the imposed environmental conditions.

1 Introduction

The climate change and other environmental factors can not only cause serious problems to human health and the surrounding ecosystem but they could also pose non-negligible impact on Civil Engineering constructions. Environmental factors could represent the climate change resulting in a meteorological drought period or the presence of vegetation, trees or extreme precipitation in a short period. These factors are crucial to be taken into account for construction or repairing purposes. One of the complex cases that deals with these environmental factors in Civil and Geotechnical Engineering are the case of residential buildings exposed to expansive clays followed by shrink-swell phenomenon which is triggered by these environmental factors. Clayey soils are mainly sensitive to drought humidification cycles resulting in shrinkage and swelling and consequently in the movements of the upper structure. In France, the losses from the impact of environmental stresses on the residential buildings are now in

J. S. McCartney and L. R. Hoyos (Eds.): GeoMEast 2018, SUCI, pp. 132–147, 2019.
https://doi.org/10.1007/978-3-030-01914-3_11

the second range of natural disasters after the floods and have costed around 5 billion euros between 1988 and 2007 (Vincent et al. 2009).

The irreversible damages due to this shrink-swell phenomenon are mostly recognized by vertical or 45-degree cracks on the building's wall which are caused by differential settlements due to the difference between the foundation soil parameters at each angle or the difference between environmental stresses at each side of the building. Figure 1 shows the schematic representation of the cyclic effect of environmental conditions (i.e. drought humidification cycles, root water uptake by trees, vegetation effect) on the expansive clays and the upper building behavior. The supporting soil tends to decrease in volume after a dry period and results in shrinkage of the supporting soil (Fig. 1(a)). It should be noted that the root water uptake can intensify the shrinkage and settlement of buildings because vegetation roots are generating more suction in dry periods. On the other hand, in Fig. 2(b) the supporting soil gains volume after a rainfall period which results in a swelling phenomenon and a complete saturated state with zero suction values. In this case the root water uptake is smaller than the drier state and would not normally intensify the movements of the soil or the building.

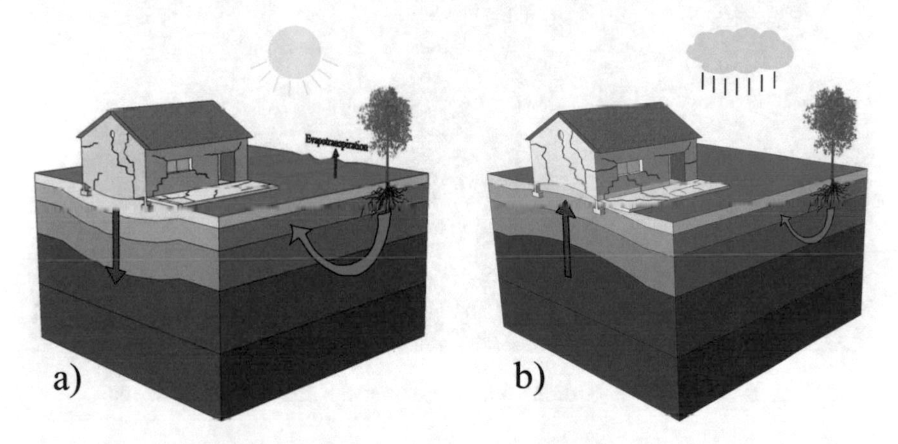

Fig. 1. The schematic representation of environmental factors affecting buildings built on expansive clays: (a) Shrinkage due to hot and dry environmental conditions (b) Swelling due to wet and humid conditions

The shrink-swell behavior of expansive clays was studied by many authors in the past years by taking into account these environmental conditions (Nowamooz et al. 2013; Adem and Vanapalli 2015; Hawkins 2013; Hemmati et al. 2012; Vincent et al. 2009; Fernandes et al. 2015; Zhang and Briaud 2015; Wray 1997; Jahangir et al. 2013; Béchade et al. 2015; Cui et al. 2013). Most of these studies dealt with the numerical simulation and the experimental behavior of expansive clays induced by different environmental stresses in laboratory scale. However, a few studies focused on the in-situ behavior of residential buildings exposed to environmental factors. Therefore, the present study investigates the effect of these factors on the behavior of the building by

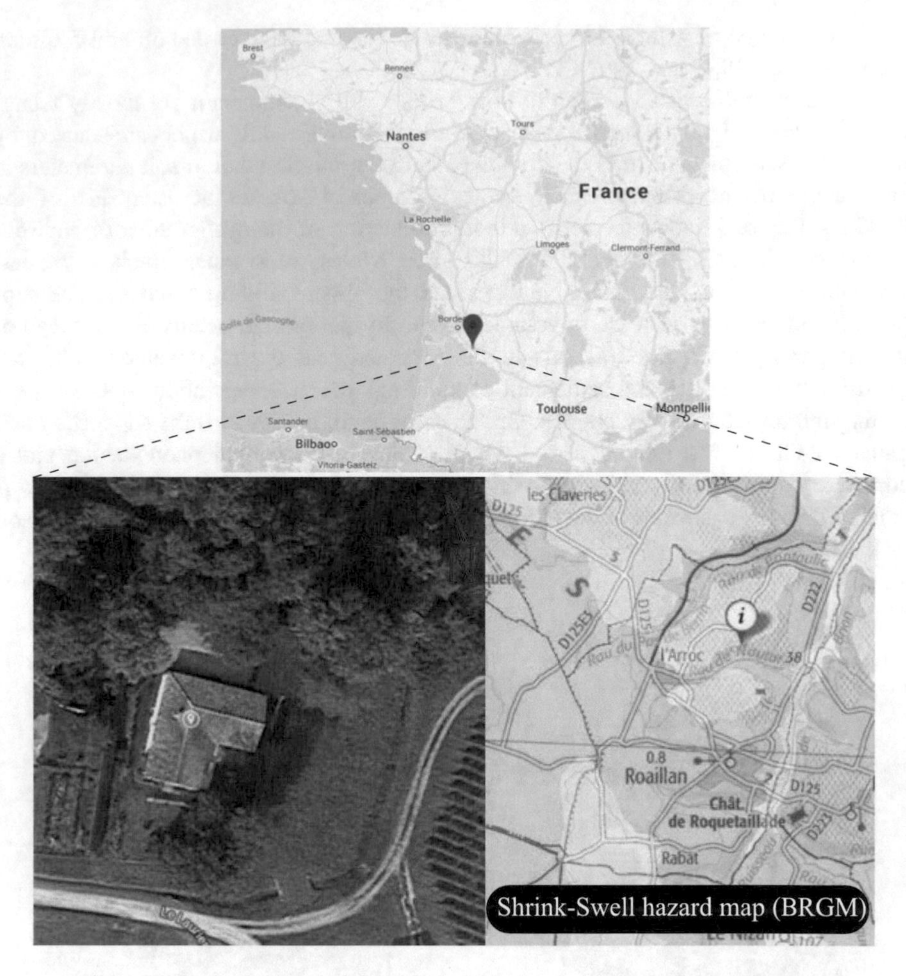

Fig. 2. Studied site location with the associated shrink-swell hazard map.

field monitoring. The site has been previously studied by four different French research organizations separately from 2011 until early 2014 however the data monitoring continued until early 2016 and the results are not published. The aim of this paper is to gather all past information and present the new results of the field monitoring program to evaluate the effect of climatic parameters and root water uptake on the structural behavior of the building.

2 Methodology

In this section, the monitoring program and the field context are primarily described. Then the climatic conditions in the site are analyzed through the soil-atmosphere interaction method.

2.1 Site Investigation

The site consists of an individual residential L shaped building and was constructed in 2005 in the south west of France with a slab on ground floor and shallow strip footings to a depth of 50 to 80 cm with a 50*30 cm section. The building was damaged due to environmental conditions in the past years followed by horizontal and vertical cracks on its walls, both outside and inside. The geological formation of the site is characterized as a medium risk of shrink swell hazard which makes the environmental factors as a highly probable cause of the observed damages (Fig. 2). Core sampling (CS) was carried out as shown in Fig. 3 at four different angles of the building to investigate the present soil state at the site in depth. Laboratory investigations on samples collected from 0.4 to 7 m depth showed that almost all soils are characterized as fine-grained and plastic. Table 1 summarizes the measured parameters based on the French soil classification system (NFP 11-300). Furthermore, these tests showed that these samples are characterized as very plastic and plastic clays. As shown in Fig. 4, the plasticity range of the samples at different angles of the building varies from intermediate to very high plasticity. In addition to that, generally the swelling potential of these samples are also characterized as highly potential using the Vijayvergiya and Ghazzaly (1973) classification which is based on the Casagrande chart. These investigations confirmed that the construction site contains mainly high plasticity expansive clays which are very sensitive to environmental conditions.

2.2 Structural Field Monitoring

In addition to meteorological conditions, the building was also exposed to the root water uptake by trees in its vicinity. Figure 3 shows the schematic representation of the building with the position of the closest trees to the building. The tree located in the north angle of the building (0.8 m diameter and 14.8 m height) is approximately 5 m away, however, the trees in the south angle of the building are located 2 and 6 m away and have a diameter of 0.65 and 1 m respectively. The building seems to be damaged mainly on the west and the south side walls which are mainly the walls being exposed to the root water uptake. In order to monitor the structural stability and the movements of the cracks, tree sides of the building were monitored since 2011 using displacement sensors installed on the observable cracks generated in 2009 (BRGM). Three sensors were primarily installed in 2011 on the south side of the building where the cracks were mostly horizontal. The measurements resolution is 0.003 mm with a precision of 0.01 mm and a $\pm$ 5 mm of measurement range followed by a data storage frequency of 8 h (i.e. 3 measurements per day). The Fissuro-Thermo-Loggers (F10TN) are able to determine if the crack is moving (closing-opening) and are also able to monitor its amplitude in a precise way (Fig. 5(a)). The configuration of these three sensors was changed in 2014 and were installed on the west side of the building where the largest crack was observed (45°) and finally, three other sensors were installed at the same time in 2014 on the east side of the building (south east angle) where vertical and 45° staircase like cracks could be observed. In addition to these displacement sensor, a settlement gauge was installed in 2014 on the north angle of the building to monitor the soil movements at different depth from 1 to 5 m. It is generally supposed that the north

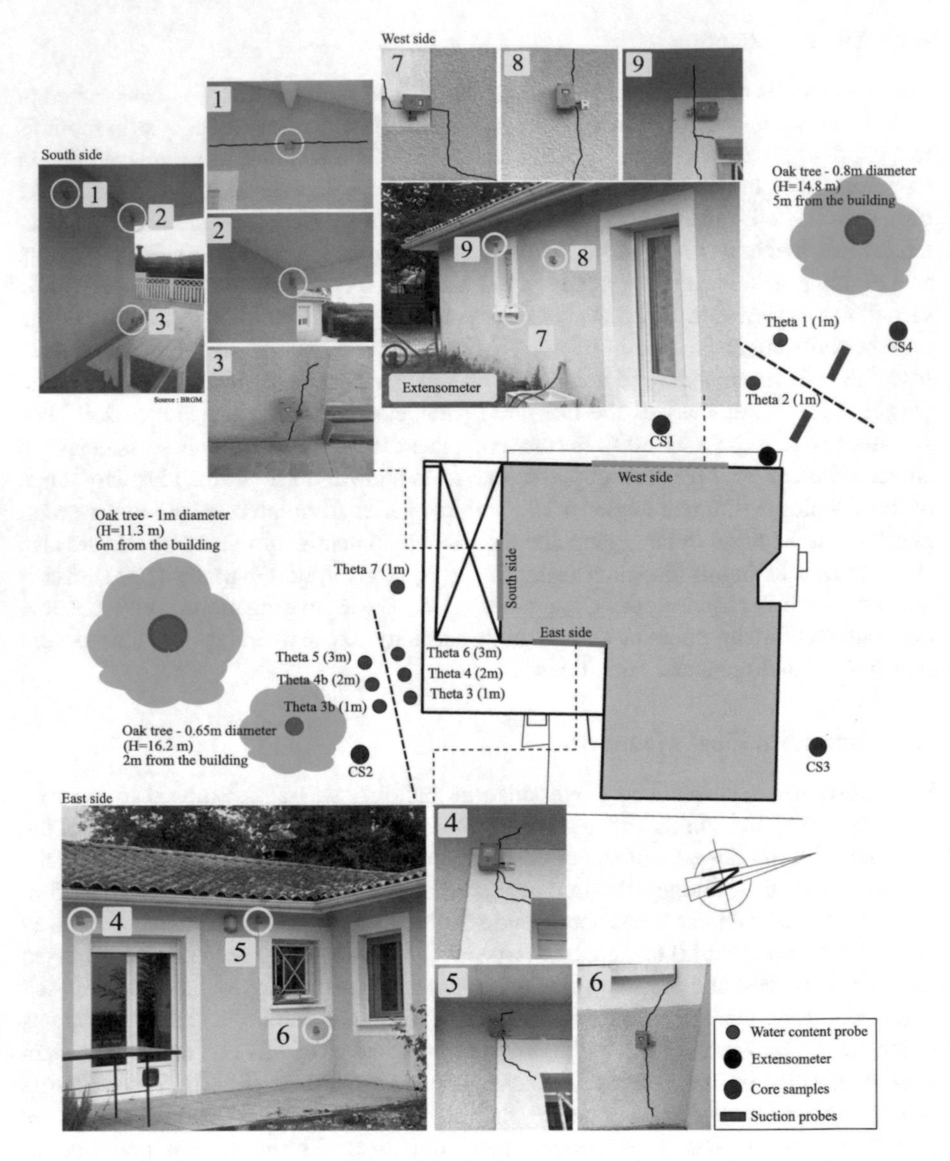

Fig. 3. The schematic representation of the building with probes and the cracked monitored walls.

angle of the building is in a more critical state compared to the south angle because of the presence of larger cracks on this side. The monitoring program is also followed by installing soil water content and suction probes which are described in the next section.

Table 1. Soil identification parameters (Mathon and Godefroy 2015).

Depth (m)	Core samples	Geological formation	∅ < 80 µm (%)	Clay particles content (%)	Liquid Limit (LL)	Plastic Limit (PL)	Plasticity index (PI)	Methylene blue value VBS	Activity (%)	GTR classification	USCS classification
0.4–0.8 m	CS1	Clayey silt	66.8	29	35	17	18	–	62.07	A2	CL
0.9–1 m	CS1	Clay	98.6	70	71	35	36	9.09	51.42	A4	MH
1.2–1.6 m	CS2	Colored clay	95.3	73	85	33	52	9.58	71.23	A4	CH
1.2–2 m	CS4	Colored clay	99.6	76	70	27	43	9.63	56.57	A4	CH
1.3–1.6 m	CS3	Colored clay	69.8	49	47	18	29	3.5	59.18	A3	CL
2.3–2.6 m	CS1	Compacted clay	99.4	–	69	30	39	7.55	–	A3	CH
2.4–2.8 m	CS3	Stiff clay + sand	93.1	75	70	29	41	9.01	54.66	A4	CH
3.1–3.2 m	CS2	Colored clay	99.8	–	68	27	41	7.95	–	A4	CH
3.6–3.8 m	CS1	Compacted clay	99.2	80	65	29	36	8.11	45	A4	CH
3.8–4.4 m	CS4	Stiff Ocher clay	99.9	79	60	26	34	9.8	43.04	A4	CH
4.2–4.3 m	CS2	Stiff clay + limestone	87.6	–	49	24	25	6.71	–	A2	CL
4.4–5.2 m	CS4	Stiff beige clay	99.1	62	47	18	29	5.51	46.77	A2	CL
5.2–5.6 m	CS1	Compacted clay	96.5	63	45	22	23	5.14	36.5	A2	CL

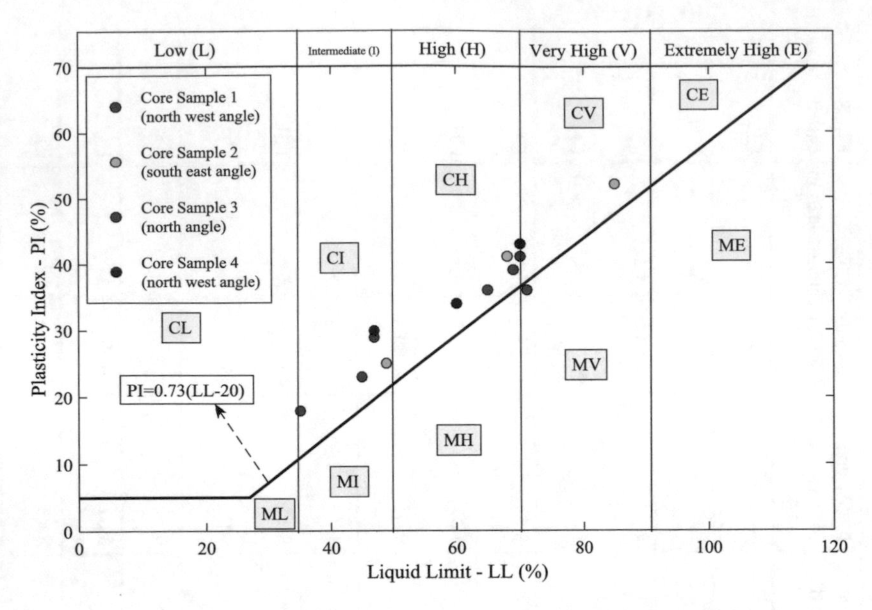

Fig. 4. Plasticity range of studied samples by the Casagrande plasticity chart.

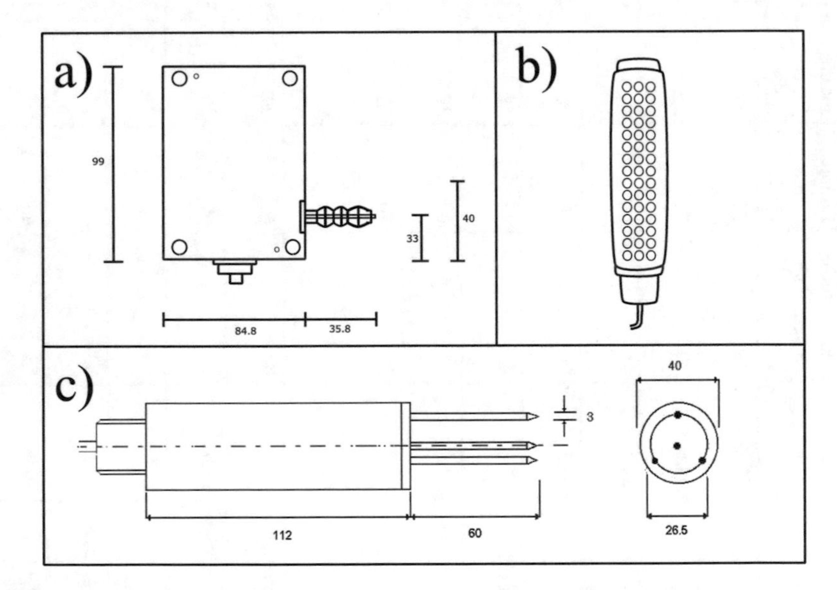

Fig. 5. Instruments used in for monitoring: (a) displacement sensor, measuring crack closing/openings (F10TN); (b) Soil matrix suction sensor (Watermark); (c) Soil volumetric water content/moisture sensor (ThetaProbe)

2.3 In Situ Soil Monitoring

As indicated in Fig. 3, suction and water content probes were installed in the north and south angle of the building at different depth. Some of these sensors were installed near the building and some of them near the trees to quantify the effect of the climatic conditions versus the root water uptake.

Soil Matrix Suction Probes

The north side of the building contains suction probes (Time Domain Reflectometry - TDR) at different depth (from 0.8 m to 3 m) which were installed by the CEREMA and were reported in Mathon and Godefroy (2015) covering the period of 2011 until early 2014. These probes were installed with an initial span of 0.75 m from the building and 0.25 m for others. The monitoring of the soil suction continued until late 2015 and is reported here. Furthermore, new suction probes were installed in 2013 starting form a depth of 2.5 m to 5 m, both close to the tree and close to the building in the north angle for monitoring the root water uptake and its effect on the soil in the building's vicinity.

Watermark type soil matrix potential sensors were used to measure the soil matric suction changes over time (Fig. 5(b)). As the soil matrix potential changes with water content, the resistance changes as well. That resistance can be measured using the Watermark Sensor. The sensor consists of a pair of highly corrosion resistant electrodes that are embedded within a granular matrix. A current is applied to the sensor to obtain a resistance value. The sensor correlates then the resistance to Centibars (kilopascals) of soil water tension.

The sensor was calibrated to obtain a relationship between the measured matrix potential (suction) and the soil water content. These measurements were carried out on 3 samples at three different depths (2.3 m and 4.3 m on core sample 1 – CS1 and at 1.8 m on core sample 3 – CS3). The following steps were adapted in the calibration process: The samples were first saturated, then the probes were installed on the samples, the samples were dried and the matrix suction was measured each 15 min. Figure 6 shows the calibrated retention curves of these three samples. These curves allow the deduction of the soil water content corresponding to the measured soil matrix suction in site. Note that the mathematical relationship can be both obtained by fitting a polynomial equation on the data or retention models in the literature.

Soil Moisture Sensors

On the other hand, seven water content probes (Theta probe) are installed in the south angle of the building in 1, 2 and 3 m depth as shown in Fig. 3. Two water content probes are installed in the north angle of the building (close to the tree and close to the building) in 1 m depth. The results of the water content monitoring for the period of 2011 until late 2013 are reported by the BRGM. However, the monitoring continued until March 2014. Therefore, the additional results of the water content probes are also reported here. The mentioned periods cover the displacement monitoring period of cracks which started in early 2014.

ThetaProbe type sensors (Fig. 5(c)) were used to measures volumetric soil moisture content, θ_v, by the well-established method of responding to changes in the apparent dielectric constant. These changes are converted into a dc voltage, virtually proportional to soil moisture content over a wide working range. These sensors measure soil parameters by applying a 100 MHz signal via a specially designed transmission line whose impedance is changed as the impedance of the soil changes. The difference between the voltage is used by the sensor to measure the apparent dielectric constant of the soil.

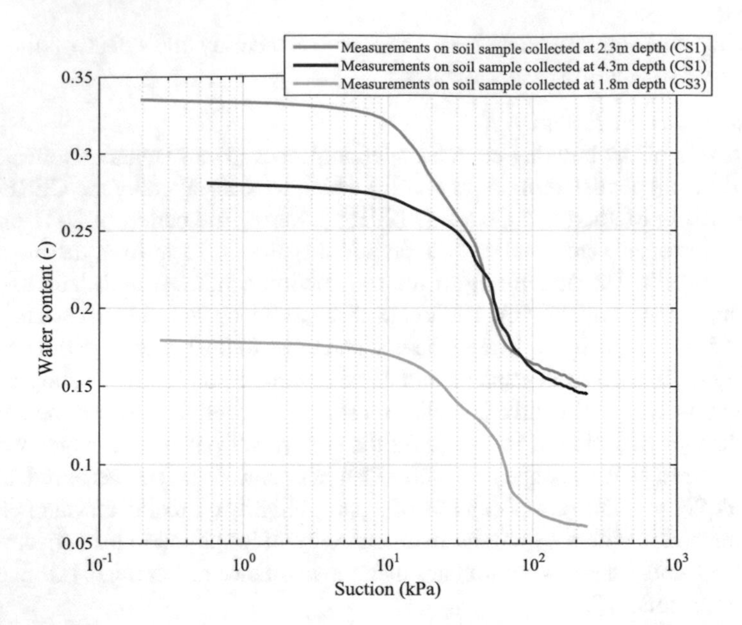

Fig. 6. Calibration of the soil matrix suction sensors with the water content for three collected samples.

Both of these sensors (Watermark and ThetaProbe) were installed by digging separate access holes for each depth which ensures that each probe is installed into undisturbed soil at the bottom of its own hole. It should be mentioned that there was no water flow into the refilled holes, however, even a failure on a single hole doesn't jeopardize all the data, as it would if all the measurements were made in a single hole. The main drawback to this method is that a hole must be dug for each depth in the profile.

2.4 Climatic Conditions

In order to assess the effect of climatic conditions on the site, soil-atmosphere interaction methods are used. This concept is based on the energy balance and the mass balance approach which is developed by Blight (1997) and gives the natural conditions at the soil surface. The simplified mass balance approach by neglecting the amount of runoff and interception in the calculations, is expressed as follow:

$$P - ET = I_{inf} \tag{1}$$

where P is the amount of precipitation (available from climatic data), ET is the amount of evapotranspiration and I_{inf} represents the rate of infiltration into the soil. From a Geotechnical Engineering perspective, these parameters are the most important ones that should be taken into account in soil-atmosphere analysis. The evapotranspiration rate can be described using the FAO 56 (Penman 1948) model as follow:

$$\lambda ET = \frac{\Delta(R_n - G) + \rho_a c_p \left(\frac{e_s - e_a}{r_a}\right)}{\left[\Delta + \gamma\left(1 + \frac{r_s}{r_a}\right)\right]} \tag{2}$$

where R_n (W/m^2) is the net radiation flux, G (W/m^2) is the soil heat flux, (e_s - e_a) is the vapour pressure deficit of the air, λ is the energy of vaporization (J/g), ρ_a is the mean air density at constant pressure, c_p is the specific heat of the air, Δ represents the slope of the saturation vapour pressure temperature relationship, γ is the psychrometric constant that relates the partial pressure of water in air to the air temperature and is equal to 66 (Pa/°C), r_s and r_a are the (bulk) surface and aerodynamic resistances that depend on the wind speed.

The energy balance approach can be used to determine the soil surface heat flux or temperature. Limited measurements of soil temperature were carried out on the site (only in one angle at two depth), therefore, the soil temperature variations are not reported here. However, the energy balance is used to determine the soil surface temperature by presenting the two terms of the energy balance equation (G, Le) in function of the net solar radiation (Rn). The energy balance equation is expressed as below:

$$R_n = G + H + L_e \tag{3}$$

Using the introduced concept of soil-atmosphere interaction analysis and the climatic data of the site (Precipitation, air temperature, wind speed, relative humidity, global solar radiation), the surface climatic conditions are deduced. The meteorological data were collected from a nearby station which provided the necessary data for the calculation of the evapotranspiration. Figure 7 presents the amount of the infiltration rate deduced by calculating the evapotranspiration and using the mass balance for the period of 09/2011 to 09/2017. The soil surface temperature is also deduced and shown along with the air temperature for the same period. The deduced climatic conditions are

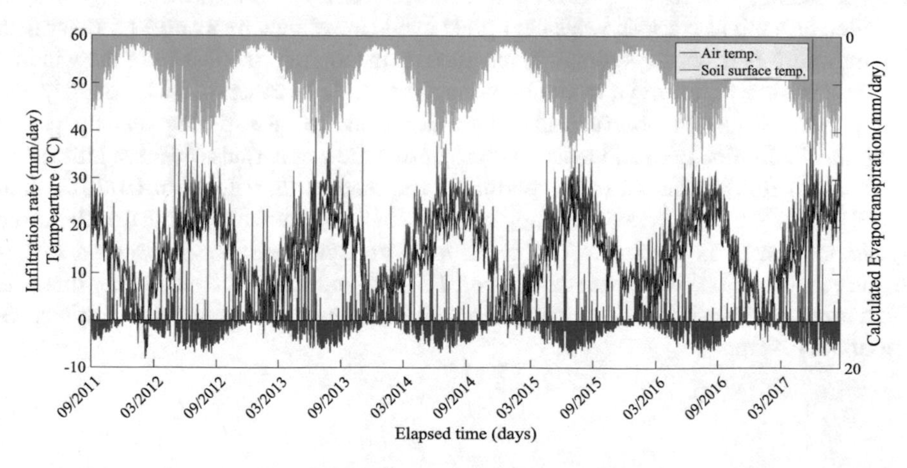

Fig. 7. Deduced climatic conditions on the site from 09/2011 until 09/2017.

used for the comparison with the soil and the structural monitoring results. Concerning soil monitoring, probes that are installed near the soil surface are much more influenced by climatic conditions, however probes installed in depth are more influenced by root water uptake if they are installed near the tree.

3 Results

In this section the results of the monitoring in the south and north angle of the building are presented.

3.1 South Angle

Figure 8 shows the obtained results from the monitoring at the south angle of the building. Figure 8(a) shows the water content variation near the building at 1, 2 and 3 m depth. It is observed that the water content variations follow the climatic conditions imposed at the site mostly at 1 and 2 m depth. On the other hand, Fig. 8(b) shows the variations of water content near the tree at 1, 2 and 3 m depth. It is observed that the variations of the water content are lower than the ones near the building and the soil near the tree shows generally higher saturated state.

Figure 8(c) shows the crack openings at the south side of the building (09/2011–08/2013). It can be observed that the sensor 1 (crack 1) located on a horizontal crack shows a stable behavior and does not move during the monitored period. The same behavior is observed for the sensor 3 (crack 3) which was located on a 45° crack at the base of the wall. The sensor 2 (crack 2) located on the beam shows cyclic movements during the monitoring, however, the range of variation is negligible. Following the stable behavior of the south side, sensors were installed on the south east wall in 01/2014 where major cracks were observed.

Figure 8(d) shows crack movements results on the south east side. Sensors 6 (crack 6) located on a crack show a complete stable state without any movement during the monitoring (01/2014–01/2015 and 08/2015–02/2016). The sensor 5 (crack 5) installed on a micro crack shows a negligible cyclic movement by a range of lower than 0.1 mm. However, sensor 4 (crack 4) installed on the top right-hand side of the window on the east side wall show a cyclic movement which is in coherence with the dry and wet periods. It can be observed that the crack tends to open during the dry period (negative infiltration rate) and tends to close during the wet periods (positive infiltration rate) which follows the climatic conditions imposed at the site from 01/2014 until 01/2015. For the next period starting from 08/2015 and ending at 02/2016, the same cyclic movement is observed. The crack tends to open until the infiltration rate is negative (dry period) which continues until 10/2015 and it tends to close after this date by starting positive infiltration rate. The maximum range of movements on this crack are about 0.4 mm.

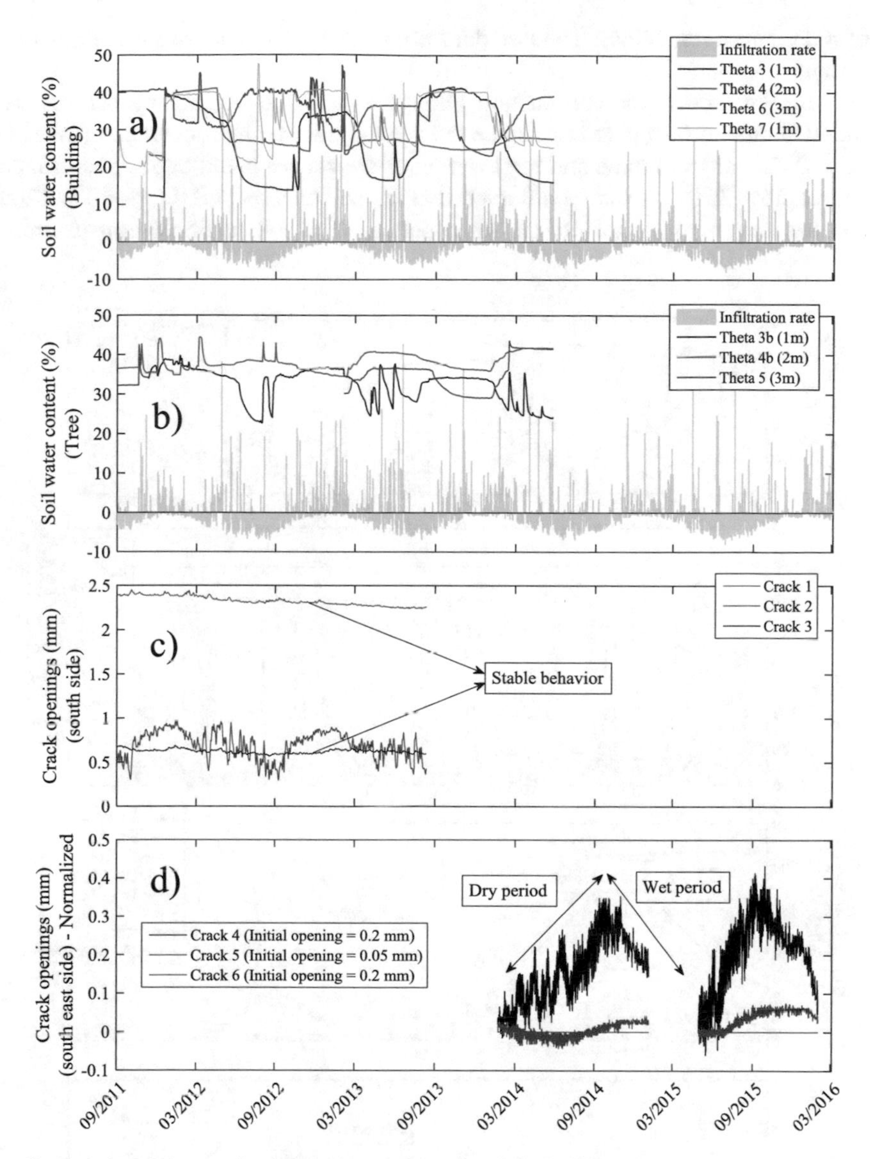

Fig. 8. Results of the monitoring at the south angle of the building: (a) water content variations near the building; (b) water content variations near the tree at three different depths; (c) crack closing/openings on the south side of the building; (d) crack closing/openings at the east side of the building

3.2 North Angle

Figure 9 show the results of the monitoring at the north angle of the building. Figure 9(a) shows the water content variations at surface (1 m) close to the tree and the building. It is observed that both probes show similar results and follow both the imposed climatic conditions. The tree does not seem to affect the soil water content at 1 m depth.

Figure 9(b) shows the soil suction variations near the tree. Many sensors were installed at different depth to investigate the suction induced by root water uptake. The details of the vertical profile and the location of the probes could be found in Mathon and Godefroy (2015). It can be observed that the soil suction at 0.8 m for T1, T2 and T3 probes shows approximately the same pattern. The soil suction seems to drop to

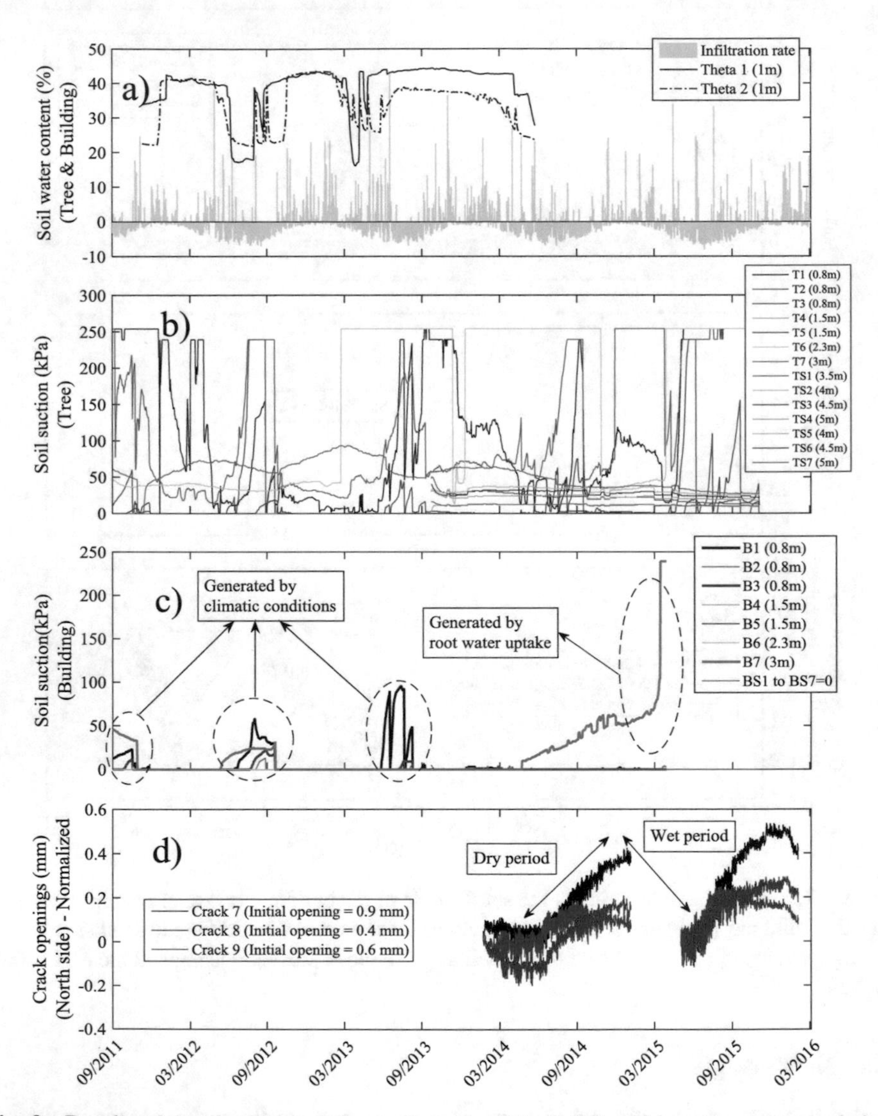

Fig. 9. Results of the monitoring at the north angle of the building: (a) water content variations near the building vs variations near the tree; (b) soil suction variation at different depth near the tree – sensors installed in the same depth have a 0.25 to 0.5 m span; (c) soil suction variation at different depth near the building; (d) crack openings at the west side of the building

almost zero at 1.5 m showing a saturated state however it shows higher values up to 250 kPa at 2.3 m. The additional sensors installed in late 2013 at 4 m, 4.5 m and 5 m depth show stable values around 40 kPa suction. It can be deduced that the desiccation front generated by root water uptake is located from 0 to 0.8 m depth and from 2.3 m to 4 m depth. The soil stays in an almost saturated state between these limits (from 1 to 2 m depth approximately) and is not influenced by the root water uptake. On the other hand, Fig. 9(c) shows soil suction variations at different depth near the building. It can be observed that the soil suction follows the imposed climatic conditions imposed on the site and shows a complete saturated state when the infiltration rate is positive. It should be noted that the values of suction are much smaller than in Fig. 9(b) which confirms that the soil near the building is not influenced by the tree. The soil suction in B1, B2 and B3 measured at 0.8 m, shows the same pattern and almost the same values. Furthermore, the B4, B5 and B6 probes showed zero suction values during the monitoring (at 1.5 m and 2.3 m) indicating the presence of a saturated phase. The B7 probe at 3 m depth shows the same results as the previous sensors until 03/2014 but it shows a sudden increase in suction values in 03/2015 (started from 03/2014) which is related to the desiccation front generated by root water uptake as shown in Fig. 9(b). It is worth mentioning that the additional sensors installed in 2013 at higher depth showed a complete saturated state of the soil with zero suction values.

Figure 9(d) shows the crack openings on the west side wall (north angle). Sensors were installed on the largest cracks observed on the building. All three sensors show cyclic movements of the crack. Sensor 7 (crack 7) with an initial opening of 0.9 mm shows larger variations than the other ones. It follows not only the climatic conditions imposed on the site, but also the root water uptake generated near the building at 3 m depth. The crack tends to open when the soil near the building shows a dry state from 06/2014 to 10/2014 and from 06/2015 to 10/2015. However, the crack tends to close in wet periods which could be observed from the data beginning from 10/2015 until 02/2016. Furthermore, the results of the extensometer installed in 2014 (not reported here) confirmed the shrink swell characteristic of the soil and consequently the crack openings.

4 Conclusions

To evaluate the effect of environmental factors on the stability of Civil and Geotechnical Engineering constructions, a damaged residential building constructed on expansive clays and exposed to environmental conditions was instrumented. Water content probes, suction probes and displacement sensors were used to study the site. The climatic conditions of the site were deduced by soil-atmosphere interaction method and were used to analyze the results obtained from the sensors. The following conclusions can be drawn from this study:

1- The differential openings of cracks and the soil physical properties on each angle show that the building is exposed to differential soil movements.
2- The soil surface (top layer) is not influenced by root water uptake in both angles of the building either close to the building or close to the tree which is confirmed by water content measurements.

3- The desiccation front on the north angle of the building located from 2.3 to 4 m can influence the soil in the vicinity of the building which is confirmed by soil suction measurements near the building.
4- The climatic conditions are dominant on the south angle. The building is not completely influenced by root water uptake which is confirmed by water content and crack opening measurements (to a depth of 3 m). The south side and south east side wall mostly show a stable behavior where the sensor 4 (crack 4) follows the deduced climatic conditions of the site.
5- Both the root water uptake and climatic conditions affect the north angle of the building. The crack openings on the west side wall are influenced by root water uptake in dry periods which is confirmed by suction measurements and they also follow the climatic conditions.

As a perspective of this study it would be interesting to monitor the soil movements at different angles of the building before and after the repairing solution of the damages in order to evaluate the performance of the adapted repairing approaches and the stability of the building.

Acknowledgments. This research was funded and supported by the French National Agency of Technological Research (ANRT) as an industrial based R&D program (CIFRE) on drought and climate change effect on natural clayey soils interaction with constructions. The authors would like to thank the IFSTTAR (The French institute of science and technology for transport, spatial planning, development and networks), the CEREMA (Center for Studies and Expertise on Risks, the Environment, Mobility and Development), the BRGM (French Geological Survey), the University of Bordeaux (I2 M Laboratory) and Alain Franck Béchade for their support on collecting in situ field data and their collaboration in this research project.

References

Adem, H.H., Vanapalli, S.K.: Soil-environment interactions modeling for expansive soils. Environ. Geotech. **3**(3), 178–187 (2015). https://doi.org/10.1680/envgeo.13.00089

Béchade, A-F., Fabre, R., Mathon, D.: Vegetation, clays and constructions. Symposium International SEC. Marne la Vallée – France (2015)

Blight, G.E.: Interactions between the atmosphere and the earth. Géotechnique **47**, 715–767 (1997). https://doi.org/10.1680/geot.1997.47.4.713

Cui, Y.J., Ta, A.N., Hemmati, S., Tang, A.M., Gatmiri, B.: Experimental and numerical investigation of soil–atmosphere interaction. Eng. Geol. **165**, 20–28 (2013). https://doi.org/10.1016/j.enggeo.2012.03.018

Fernandes, M., Denis, A., Fabre, R., Lataste, J.F., Chretien, M.: In situ study of the shrinkage of a clay soil cover over several cycles of drought-rewetting. Eng. Geol. **192**, 63–75 (2015). https://doi.org/10.1016/j.enggeo.2015.03.017

Hemmati, S., Gatmiri, B., Cui, Y.J., Vincent, M.: Thermo-hydro-mechanical modelling of soil settlements induced by soil-vegetation-atmosphere interactions. Eng. Geol. **139–140**, 1–16 (2012). https://doi.org/10.1016/j.enggeo.2012.04.003

Jahangir, E., Deck, O., Masrouri, F.: An analytical model of soil interaction with swelling soil during droughts. Comput. Geotech. **54**, 16–32 (2013). https://doi.org/10.1016/j.compgeo.2013.05.009

Mathon, D., Godefroy, A.: Monitoring of an instrumented house damaged by drought. Symposium International SEC. Marne la Vallée – France (2015)
Nowamooz, H., Jahangir, E., Masrouri, F.: Volume change behavior of a swelling soil compacted at different initial states. Eng. Geol. **153**, 25–34 (2013). https://doi.org/10.1016/j.enggeo.2012.11.010
Penman, H.L.: Natural evaporation from open water, bare soil and grass. Proced. R. Soc. Lond. Ser. A **193**, 120–145 (1948). https://doi.org/10.1098/rspa.1948.0037
Vijayvergiya, V.N., Ghzzaly, O.I.: Prediction of swelling potential for natural clays. In: Proceedings of the 3rd International Conference on Expansive Soils, vol. 1, Haïfa, pp. 227–236 (1973)
Vincent, M., et al.: Projet ARGIC—Analyse du Retrait Gonflement et de ses Incidences sur les Constructions. Projet ANR-05-PRGCU-005. Rapport final, Rapport BRGM/RP-57011 13 pp. et 39 annexes (2009)
Wray, W.K.: Using soil suction to estimate differential soil shrink or heave. Unsaturated soil engineering practice, geotechnical special publication 68, pp. 66–87. American Society of Civil Engineers, Reston, Virginia, USA (1997)
Zhang, X., Briaud, J.L.: Three dimensional numerical simulation of residential building on shrink–swell soils in response to climatic conditions. Int. J. Numer. Anal. Methods Geomech. **39**, 1369–1409 (2015). https://doi.org/10.1002/nag.2360

Monitoring the Efficiency of Polyurethane Resin Injection for Foundation Remediation in Damaged Residential Buildings Exposed to Expansive Clays

Hossein Assadollahi[1,3](✉), L. K. Sharma[2], Anh Quan Dinh[3], and Blandine Tharaud[4]

[1] Department of Civil Engineering and Energies, ICUBE, UMR7357, CNRS, INSA, 67000 Strasbourg, France
hossein.assadollahi@insa-strasbourg.fr
[2] Department of Earth Sciences, Indian Institute of Technology, Bombay, India
[3] Department of Engineering and Consulting, DETERMINANT R&D SARL, 75008 Paris, France
[4] Department of Geotechnical Engineering and Investigation, DETERMINANT SARL - Solinjection, 13590 Meyreuil, France

Abstract. This paper deals with the monitoring of the efficiency of polyurethane resin injection as a treatment technique in damaged residential buildings. The polyurethane resin technique is mainly adapted in repairing shallow foundations by densifying the soil and increasing the bearing capacity. The distribution path of this expansive type of material into the soil is not always known which makes it difficult to avoid further structural damage to the building. To cope with this issue, a field measurement technique was adapted for a case of a damaged residential building due to the presence of shrink-swell clays under the foundation. A geotechnical investigation survey was carried out both at laboratory and field scale before and after the treatment process. Physical properties of the soil were investigated before and after the treatment process in the laboratory and the in-situ soil dynamic resistance was investigated before and after the treatment process by the PANDA test (dynamic penetrometer test). During the treatment process, four optical fiber sensors were installed on four angles of the building in order to measure the local displacements of the walls and the cracks. Results showed that the real time monitoring of the building is an efficient technique to avoid additional damage.

1 Introduction

In recent years, new technologies in repairing damaged Civil Engineering constructions have been the subject of study and research in different areas. Lightweight Civil Engineering constructions like residential buildings, industrial buildings or even roads are easily damaged when their supporting soils are subjected to severe hydro-meteorological drought events. These drought periods can affect the supporting soils, especially constructions built directly in contact with clayey soils, by decreasing the

J. S. McCartney and L. R. Hoyos (Eds.): GeoMEast 2018, SUCI, pp. 148–157, 2019.
https://doi.org/10.1007/978-3-030-01914-3_12

bearing capacity due to different climatic cycles. Clayey soils can easily shrink and swell under hot and humid climatic conditions generating differential settlements on the soil and in consequence causing structural damage on lightweight constructions (Jahangir 2011). This phenomenon has been studied by different authors in the past years by taking into account coupled and uncoupled behavior of clays interacting with the atmosphere or in contact with building constructions (Hemmati et al. 2012; Adem et al. 2013; Fernandes et al. 2015). Constructions with shallow foundations are much more sensitive to this phenomenon and in most cases, are not able to properly transfer the upper loads to the ground which results in structural damages. Cracks are common problems affecting the stability of buildings if they are not treated quickly. Not only structural cracks on buildings can cause problems but also, foundation soils are potential factors to be taken into account. Usually cracks are sealed with special mortars or structural elements are reinforced, if the cause of damages are not related to foundation soil, however, if the cause of damages is related to the foundation soil, Geotechnical repairing techniques should be adapted before repairing the cracks. These damaged constructions can be repaired by different techniques, one of the well-known and traditional methods is the execution of micro piles that is executed on the existing foundation in order to increase its bearing capacity. Although this method is reliable for geotechnical engineers, but it is not always cost effective and can be very expensive in some cases and the way they are executed on the site is sometimes complicated and time consuming. The alternative method recently adapted and developed for repairing damaged and cracked buildings, is the injection of expansive polyurethane resin into the supporting soil. The injected resin, densifies the soil by expanding itself and increases the soil density and bearing capacity. Although this method is much more cost effective and is not time consuming but the evaluation of its efficiency stays a big challenge. Many authors showed that it could be very efficient in laboratory scale by studying its physical and mechanical characteristics (Buzzi et al. 2008; Buzzi et al. 2010; Valentino et al. 2014; Svaldi et al. 2005). Santarato et al. (2011) investigated its expansion into the soil by using 3D electrical tomography technique. Nowamooz (2016) investigated its expansion using numerical methods. However, these methods do not deal with the real time displacements of the building due to the resin injection technique and the soils characteristics improvements at the same time. Lightweight constructions like residential buildings are very sensitive to the injection pressure, so that a higher amount of injection can cause irreversible damage on the structure by lifting it upward. To avoid such problems, in situ monitoring is proposed to be executed on the structure. The schematic representation of the procedure is shown in Fig. 1 indicating that the effect of the injection process on the structure's cracks is monitored by fiber optic sensors, while the soil is densifying in order to create an alerting system to avoid additional displacements of the building. In addition to that, it is crucial to investigate the obtained improvements in the soil's in situ mechanical characteristics. Therefore, this paper presents the results from in situ measurements and monitoring of a damaged residential building, repaired by the polyurethane resin injection technique.

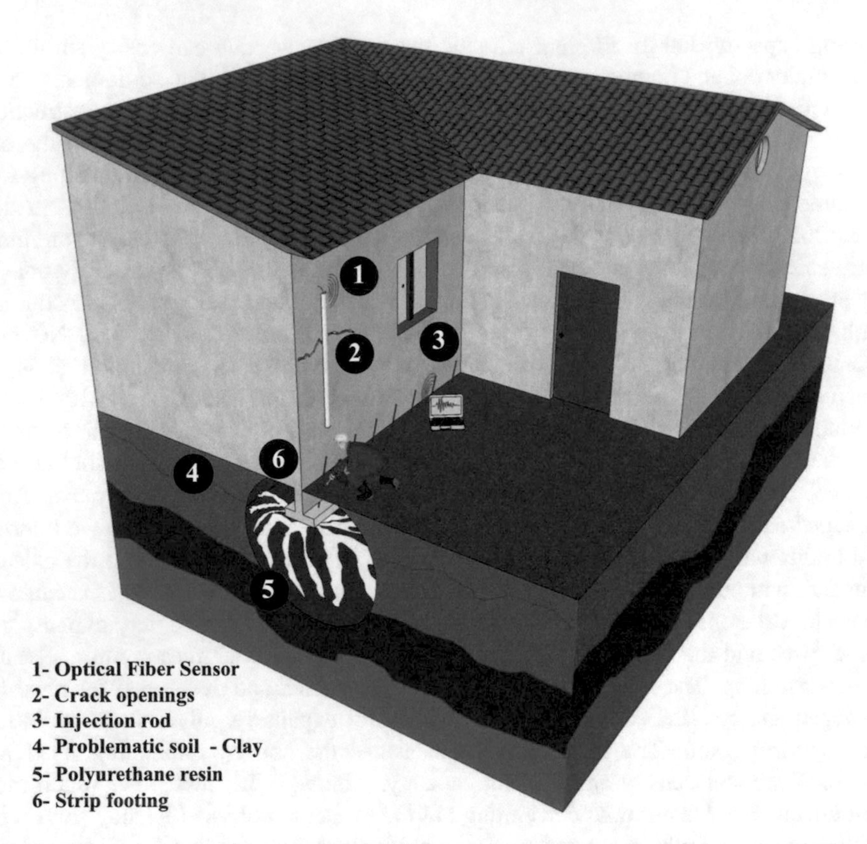

Fig. 1. The general monitoring process and foundation soil treatment of damaged buildings.

2 Methodology

2.1 Site and Laboratory Investigation

The studied site concerns a one floor residential building constructed with strip footing which is damaged by 45° and vertical cracks on its walls (Fig. 2). The building is located in the west side of the Castelnau d'Estretefond region. The geological data provided by the French Geological Survey (BRGM) map, classify the site to be composed of modern Alluviums, constituted by silts and gravels. To monitor the improvements of the soil characteristics, Geotechnical investigations were carried out before and after the injection process both at laboratory and field scale. Four core samples were first collected and studied in laboratory (two from 1 to 5 m depth and two others from 1 to 3 m depth). The natural water content was measured as shown in Table 1 (before and after injection). The swelling potential of the specimen at 2 m depth (under the foundation) was classified as high swelling capacity depending on the soil samples Atterberg's limit (Table 2). In addition to the soil laboratory investigation, in situ PANDA tests were carried out in order to determine the dynamic tip resistance of the soil in depth at different points around the building allowing to have a general idea of the mechanical characteristics of the geological formation before and after treatment. The concept of a PANDA test is to drive a cone fixed at the end of a set of rods into the soil using a hammer. The depth of rod penetration and the tip dynamic resistance q_d are recorded automatically after each hammer hit. The dynamic tip resistance is expressed as below:

$$q_d = \frac{M_g H}{(1 + a)e} \tag{1}$$

where M is the hammer mass, H is its falling height, a is the ratio of masses (a = P/M, the rod-system penetrated mass, P, over the hammer mass, M), e is the penetration of the rod after impact and g = 9.8 m/s^2. Core samples and PANDA tests points are presented in Fig. 2.

2.2 Monitoring Procedure

In addition to the monitoring of the soil characteristic, it is of great importance to monitor also the displacements of the building in real time. It is important to identify the damaged areas on the building, especially cracks openings before any monitoring procedure. The expansive polyurethane resin was injected under the foundations and some parts of the floor slabs mainly on the north side of the building as a repairing solution. The important point in the process is to be as efficient as possible in order to avoid any secondary damage on the structure. This could be done by monitoring the upper structure displacements. In order to monitor the injection process in real time, optical fiber sensors were used. Four optical fiber sensors were installed on four different angles of the building (Fig. 2) which were used in dynamic mode with measurement time steps at each 20 ms with a resolution of 20 μm. Each strain measurement was made when the injection was taking place at that specific point. Three sensors were installed on the building's walls on very small cracks however, one of the sensors (sensor 4) was installed on the largest crack of the building. Each of these measurements points were followed by a complementary PANDA test to evaluate the dynamic resistance profile after the treatment phase.

3 Results and Discussion

Results of the laboratory experiments on the samples before and after the injection (1/1*) show that the plasticity index and the swelling potential decreases after the injection phase. The water content measurements in the core samples 1/1* (Table 2) tend to increase after the injection phase while it should normally decrease or at least stay constant because the temperature tends to increase in the injection process. On the other hand, the water content measurements in core samples 2/2* (Table 1) show that the water content has significantly decreased. It should be noted that the soil heterogeneity and the resin distribution in the material is not always homogenous and samples may not be affected by the injection at all. Furthermore, results of the in-situ

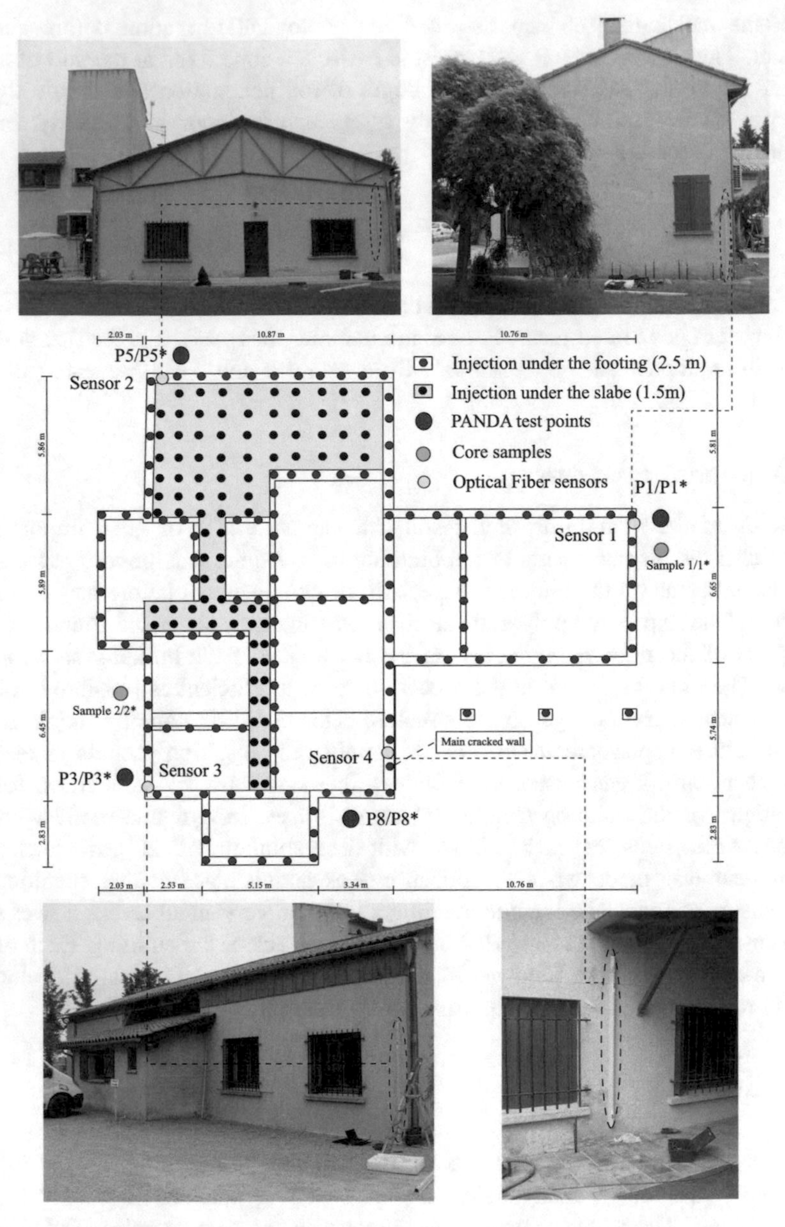

Fig. 2. The general map of the studied building and the position of geotechnical test points with the location of the optical fiber sensors.

monitoring by fiber optics along with the dynamic resistance of the soil are presented in Fig. 3. Each of these four measurements results are analyzed here:

- Figure 3a shows that the injection under the monitored wall with sensor 1, has caused a slight tension of the optical sensor after injecting 25 kg of resin. It showed

Table 1. Core sampling and water content measurements at different depth before and after the injection process

Before Injection				After injection			
Core sample	Depth (m)	Geological formation	Natural water content (%)	Core sample	Depth (m)	Geological formation	Natural water content (%)
1	1	Silty Clay	14.1	1*	1	Plastic silty clay	23.2
	2	Silty Clay	14		2	Plastic silty clay	17.9
	3	Dry sand	4.7		–		
	2	Dry sand	4.6		–		
	5	Dry sand	4.1		–		
2	1	Plastic Clay	23.8	2*	1	Plastic clay	14.3
	2	Silty Clay	17.9		2	Sandy silt	7.7
	3	Clayey sand	12.9		3	Silty sand	3.7

Table 2. Swelling potential of a specimen at 2 m depth before and after the injection process.

Core sample at 2 m	Natural water content (%)	Liquide Limit w_L (%)	Plastic Limit w_P (%)	Plasticity index PI (%)	Consistency index CI (%)	Classification
1	14	48.7	19.2	29.5	1.2	Very plastic clay
1*	17.9	31	16	14.9	0.9	Low to medium plastic clay

a stable behavior of the structure during the injection. The wall was slightly compressed (0.02 mm) during the injection and when the injection was completed, it went back to its initial state (generating a tension of 0.01 mm which seems linked to the evolution of the temperature during the monitoring process). The results of the PANDA test at this point (P1/P1*) show a significant improvement of the dynamic resistance in the first 60 cm of the soil profile but a decrease from 60 cm to 2 m depth. The test after the injection has stopped quickly than the one carried out before the injection. It is generally concluded that the wall at this point was rising up stably and the soil has been improved in the first meter.

- Figure 3b shows mainly a tension of the optical fiber from the beginning of the injection. 35 kg of resin was injected at this point but the results of the optical fiber measurements did not show any compression at the beginning meaning that the resin was not being distributed at the considered spot under the footings. It was then observed that the resin was diffused under the slab instead. However, a compression phase and a rising up of the wall is captured by the sensor 2 at the end of the monitoring period. The injection was stopped right after the compression phase was

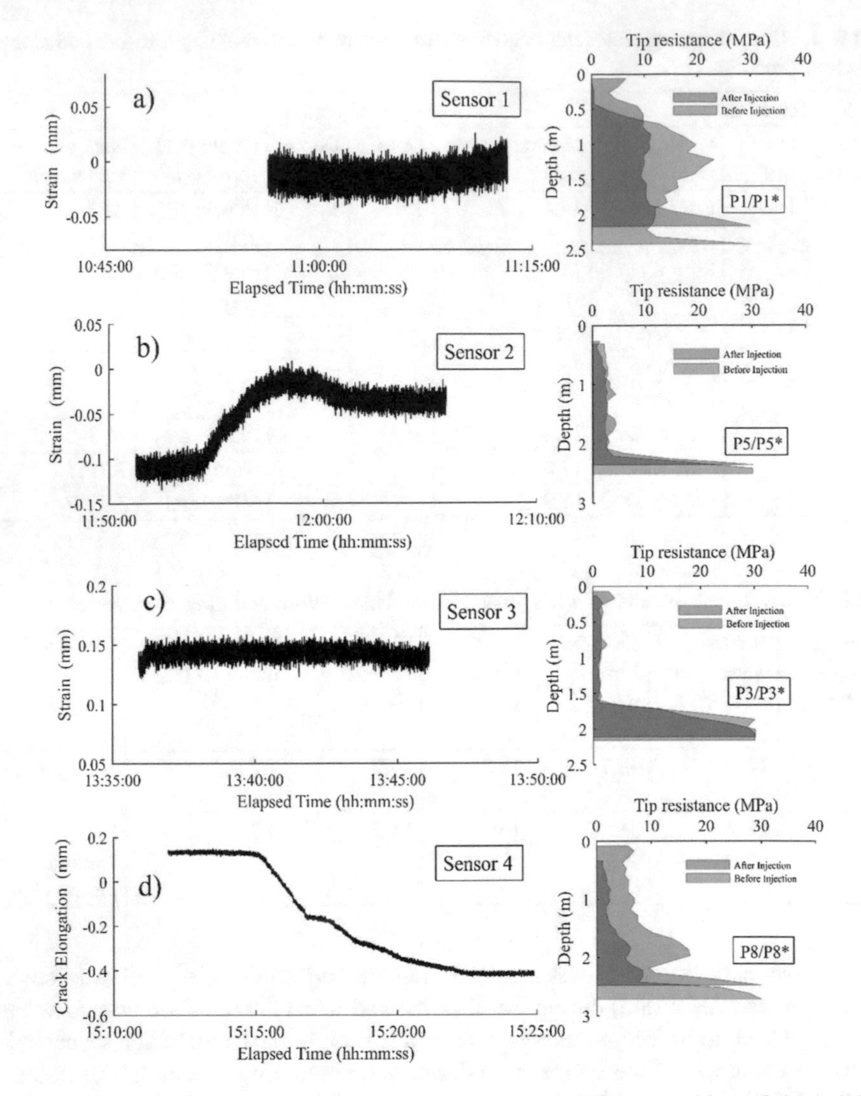

Fig. 3. Results obtained from the optical fiber sensors at 4 different position of the building along with the results of in-situ PANDA tests before and after the injection process.

observed. The results of the PANDA test at this point (P5/P5*) did not show any changes in the soil profile as expected because the resin was not distributed in the considered spot.

- Figure 3c shows that the injection under the footing at this point has led to an overall rise in the structure resulting in a stable response of the optical fiber sensor. 25 kg of resin were injected under this wall without causing any damage. Results of the PANDA test in the vicinity of this point (P3/P3*) show a slight improvement in the top layer of the profile and almost no change in the dynamic resistance after the injection phase. It can be derived that the complementary PANDA test after the injection (P3*) was not able to capture the effects of the injection but the stability of the structure in rising up the wall was confirmed with the optical fiber sensor.

- Figure 3d shows the results of the crack openings observed by the 4th optical fiber sensor installed on the largest external crack and potentially the most damaged part of the building. This unique curve shows that the resin injection under the foundation soil resulted in a closure of the crack in a range of 0.5 mm. It should be noted that the closing of the largest crack was observable with the naked eye during the injection phase. 27 kg of resin were injected at this point and the results from the PANDA test carried out before and after the injection phase (P8/P8*) show significant improvements in the whole soil profile by at least two times higher. Figure 4 describes in detail each step of the resulting strain curve of the crack. The first stage corresponds to the resin filling phase and a stable state which means that the soil is densifying. The second stage corresponds to the primarily sealing of the cracks in which the resin is taking effect and a compression of the crack is observed. The third stage shows the secondary sealing of the cracks along with a stable rising up of the structure meaning that the building tends to move upward after the crack is sealed. The last stage shows the stabilized phase in which the crack is completely sealed and the structure does not move anymore, therefore the injection phase could be stopped at this point. Without the help of monitoring, the right moment for stopping the injection process wouldn't be known which could have led to additional damage on the building.

4 Conclusions

A case of a damaged building due to the presence of shrink-swell soil was investigated and repaired by the resin injection technique. The soil laboratory investigations before and after the injection phase on a sample showed that the swelling potential of the sample decreased after the injection while the water content profile couldn't give a reasonable argument of the injection effect on the samples' physical characteristics. It should be noted that this may also be because of the heterogeneities in the in situ geological formation. The in-situ improvements could not be completely verified by laboratory investigations however, the in-situ geotechnical investigation showed much more coherent results. The use of optical fiber sensors and the in situ mechanical test of the soil dynamic tip resistance together is a combined monitoring technique that can give an idea of the strain and the movement of the structure, along the improvements of the soil mechanical parameters. For this case, sensors 1 and 3 showed a stable rising of the structures and the wall along with acceptable improvements of the soil dynamic resistance with the PANDA test mostly in the top layer. For the case of the sensor 2, the resin wasn't exactly distributed at the considered spot, so the PANDA test couldn't show the improvement at that specific point, furthermore the use of the sensor allowed the monitoring of the downward movement of the wall and was adapted as an alerting system. The sensor 4 is a unique example of a re-pairing technique in which the crack is being monitored and is being closed as the injection is taking place.

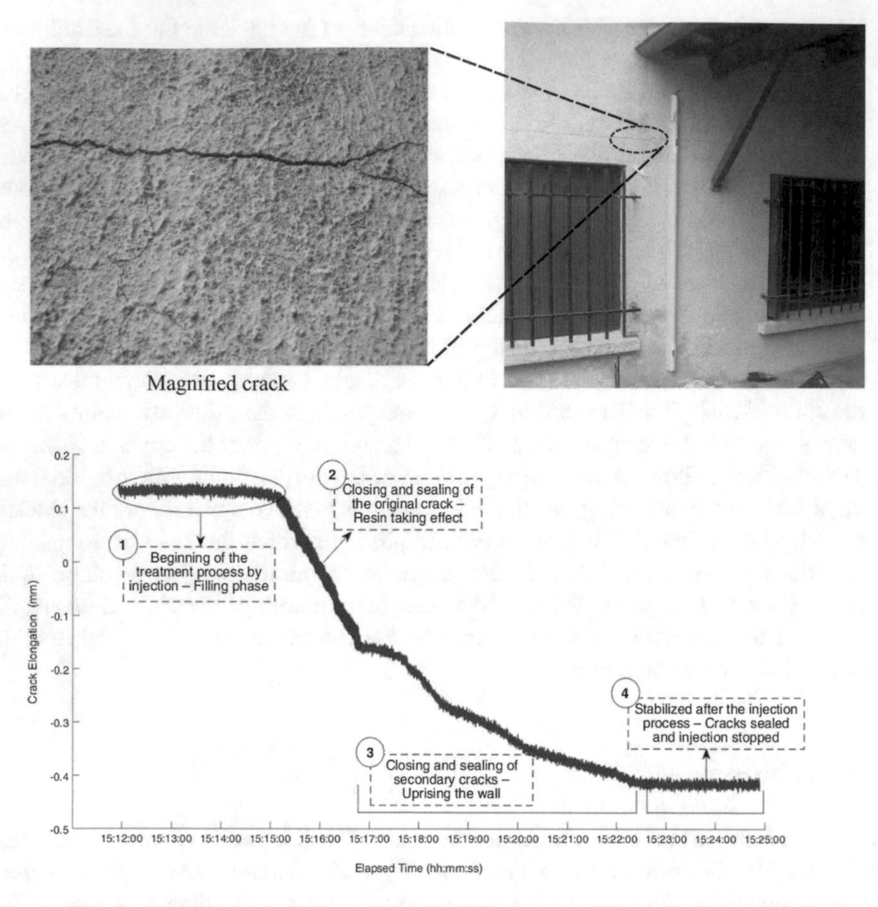

Fig. 4. Results obtained from the optical fiber sensor (4) on the largest crack.

It can be concluded that the injection can continue on until the sensors are showing stable strain values and until a tension has not taken place (meaning a downward movement). If sensors are installed on cracks, the injection should be stopped after a complete compression and a stabilization phase is shown on the sensor results. This technique could be used in further geotechnical monitoring projects in order to guarantee the safety of the structures. It should also be noted that the uncertainty in the distribution of the resin into the soil medium does not make the monitoring procedure easy and pose serious problems if the right tools are not adapted. For this case the use of optical fiber sensors is recommended however classical extensometers could also be used to measure the amount of soil movements. Further geotechnical study both in laboratory and field scale could give better understanding of this uncertainty.

Acknowledgments. The authors would like to acknowledge DETERMINANT Group for its financial support and contribution in this industrial based R&D project. This research work wouldn't be carried out without the help of site engineers and their precision in installing the monitoring system. Our special thanks go to the geotechnical site engineers for their participation in this project.

References

Adem, H.H., Vanapalli, S.K.: Soil-environment interactions modeling for expansive soils. Environ. Geotech. **3**(3), 178–187 (2015). https://doi.org/10.1680/envgeo.13.00089

Buzzi, O., Fityus, S., Sasaki, Y., Sloan, S.: Structure and properties of expanding polyurethane foam in the context of foundation remediation in expansive soil. Mech. Mater. **40**, 1012–1021 (2008). https://doi.org/10.1016/j.mechmat.2008.07.002

Buzzi, O., Fityus, S., Sloan, S.: Use of expanding polyurethane resin to remediate expansive soil foundations. Can. Geotech. J. **47**(6), 623–634 (2010). https://doi.org/10.1139/T09-132

Fernandes, M., Denis, A., Fabre, R., Lataste, J.F., Chrétien, M.: In situ study of the shrinkage of a clay soil cover over several cycles of drought-rewetting. Eng. Geol. **192**, 63–75 (2015). https://doi.org/10.1016/j.enggeo.2015.03.017

Hemmati, S., Gatmiri, B., Cui, Y.-J., Vincent, M.: Thermo-hydro-mechanical modelling of soil settlements induced by soil-vegetation-atmosphere interactions. Eng. Geol. **139–140**, 1–16 (2012). https://doi.org/10.1016/j.enggeo.2012.04.003

Jahangir, E.: Phénomènes d'interaction sol-structure vis-à-vis de l'aléa retrait-gonflement pour l'évaluation de la vulnérabilité des ouvrages. Mécanique des solides. Institut National Polytechnique de Lorraine - INPL (2011)

Nowamooz, H.: Resin injection in clays with high plasticity. C. R. Mecanique **344**, 797–806 (2016). https://doi.org/10.1016/j.crme.2016.09.001

Santarato, G., Ranieri, G., Occhi, M., Morelli, G., Fischanger, F., Gualerzi, D.: Three dimensional electrical resistivity tomography to control the injection of expanding resins for the treatment and stabilization of foundation soils. Eng. Geol. **119**(1–2), 18–30 (2011). https://doi.org/10.1016/j.enggeo.2011.01.009

Svaldi, A.D., Favaretti, M., Pasquetto, A., Vinco, G.: Analytical modelling of the soil improvement by injections of high expansion pressure resin. Bull. Angew. Geol. **10**(2), 71–81 (2005)

Valentino, R., Romeo, E., Stevanoni, D.: An experimental study on the mechanical behaviour of two polyurethane resins used for geotechnical applications. Mech. Mater. **71**, 101–113 (2014). https://doi.org/10.1016/j.mechmat.2014.01.007

Experimental Study on Distributed Optical Fiber Sensor Based on BOTDR in the Model Tests of Expansive Soil Slope Protected by HPTRM

Yingzi Xu[1(✉)], Fu Wei[2(✉)], Guang Fan[2(✉)], and Lin Li[3(✉)]

[1] Guangxi Key Laboratory of Disaster Prevention and Engineering Safety, College of Civil and Architectural Engineering, Guangxi University, Nanning 530004, China
xuyingzi@gxu.edu.cn

[2] College of Civil and Architectural Engineering, Guangxi University, Nanning 530004, China
53803184@qq.com, 757989721@qq.com

[3] Department of Civil and Environmental Engineering, Jackson State University, Jackson, MS 39217, USA
li@jsums.edu

Abstract. Brillouin optical time domain reflectometer (BOTDR) is a testing technology and monitoring technique, which utilizes Brillouin spectroscopy and optical time domain reflectometer to obtain axial strains along the distributed optical fibers. Strains can be transformed into longitudinal deformations through simple calculations. The High-Performance Turf Reinforcement Mats (HPTRM) is a kind of three-dimensional geonet, which has good extensibility and tensile performance. As the flexible material, HPTRM is applicable for expansive soil slope protection. It is a fresh attempt to deformations observation of HPTRM that are used to protect expansive soil slope by distributed fiber optic technology. The objective of this paper is to study applicability of distributed optical fiber as deformation monitoring device for expansive soil slope. In the model test, optical fibers were fixed to HTPMR by adhesive agent, and expansive soil was covered with HPTRM closely. Expansive soil was constrained by HPTRM after absorbency swell, and the HPTRM and optical fibers was subject to distortion by soil expansion. Distributed optical fiber sensor was pasted on the HPTRM to test strains of its fibers, and vertical deformation of the soil can be obtained by simple calculations. The experimental results show that the results of calculation are in good agreement with the measurement of dial indicator. And the deformation of HPTRM in protected expansive soil slope is within the allowable range of HPTRM material. It is feasible to be used in the field of deformation surveillance of HPTRM to protect expansive soil slope.

J. S. McCartney and L. R. Hoyos (Eds.): GeoMEast 2018, SUCI, pp. 158–164, 2019.
https://doi.org/10.1007/978-3-030-01914-3_13

1 Introduction

Surface prediction is an important part of expansive soil slope engineering. The anchored reinforced vegetation system (ARVS) is a new effective means of expansive soil slope protection. The ARVS consists of high-performance turf reinforcement mat (HPTRM) and percussion driven earth anchors (PDEA). HPTRM is a three-dimensional, lofty, woven polypropylene geotextile (Clifton 2010). HPTRM is made of soft and durable fabric, to produce integrated protective performance. Anti-erosion properties of HPTRM is proved by existing studies (Kang 2015). HPTRM also can reduce deformation of expansive soil slope surface (Xian 2016). These studies show that this method has many advantages in environment and protection property.

Deform action monitoring is an extremely important job in slope engineering. In traditional slope projects, dial indicators are usually used for measure soil vertical deformation. The major drawback of traditional method is that environmental factors are likely to influence the precision of the dial indicators. In recent years, distributed optical fiber sensors have been applied to deformation inspection in different engineering. This approach is superior to conventional methods in automation, integrated and less susceptible to objective factors. By distributed optical fiber sensor, the internal structural parameters of slope can be obtained (Li et al. 2008). The slope model test verified the feasibility of the distributed optical fiber system in slope engineering (Wang et al. 2009). Optical fiber materials are sensitive to temperature and require temperature correction of test data (Zhu et al. 2010). At present, the research on the application of distributed optical fiber sensor for expansive soil slope protection is limited.

In this study, verified by the self-made model test, the reliability of distributed optical fiber sensors for deformation monitoring of expansive soil protected by HPTRM is analyzed.

2 Deformation Tests of Expansive Soil Protected by HPTRM

2.1 Distributed Optical Fiber Sensor

In this study, the NBX-6050-type NEUBRESCOPE was used to obtain the axial strain of the optical fibers, which covered HPTMR. Brilliouin optical time domain reflectometer (BOTDR) is a testing technology and monitoring technique, which utilizes brillouin spectroscopy and optical time domain reflectometer to measure axial strain along the optical fibers. The strain was measured by BOTDR optical fiber sensor. The optical fibers and HPTRM extended at the same time, while the fibers fixed on the HPTRM.

Spatial resolution must be set before sensor run. It determines the size of minimum unit which is distinguishable. By choosing appropriate spatial resolution can obtain precise monitoring data and more importantly, the data processing can be simplified. NBX-6050 is set to progressive mode with a 10 cm precision.

Table 1. The performance parameters of the optical fiber

Fiber core type	Minimum bending radius cm	Maximum tension N	Elastic modulus GPa	Gauge factor C_{11} MHz/10^{-6}	Temperature coefficient C_{12} MHz/°C	Diameter mm
Single-core & Single-mode	3	160	0.2	0.04998	1.92	0.9

The optical fiber sensors in this test were made by Wuhan Senhao Communication Technology Co., Ltd. The performance parameters of the optical fiber are shown in Table 1.

2.2 Model for Expansive Soil Slope Protected by AVRS

The self-made model test apparatus was a round steel plate with baseboard on a rigid table to simulate expansive soil slope surface protected by AVRS. The geometry of the plate was 200 mm in height and 500 mm in radius. An active rigid anchor bolt was set in the center of the plate with a diameter of 100 mm anchor plate under the anchor head to press the expansive soil surface. The anchor bolt was connected to the load device through the round plate and rigid table. The round plate was filled with expansive soil and covered with HPTRM. A hoop was set at the top of the round steel plate to fasten HPTRM. As shown in Fig. 1. The grass seeds mixed with fertilizer and soil were laid on HPTRM.

The mechanical properties of HPTRM in this test are shown in Table 2. Expansive soil in this study was dark gray clay from Nanning, China. Soil properties are shown in Table 3. The results show that the free swelling rate of the sample soil was 70%, which is in the range of medium-expansive soil. The grass seeds in this test were ryegrass and Bermuda grass with the same proportion. The tests were begun after grasses were grown up.

2.3 Experiment Plan

Optical fibers and dial gauges were set on HPTRM of the model

Optical fibers were placed orthogonally on HPTRM (Fig. 2). It was fixed to HTPMR using bonding agent which mixed with epoxy resin and curing agent in a ratio of 1:2. When using optical fibers as sensing medium, the effect of temperature on optical fibers strain cannot be ignored. In order to correct the deviation caused by temperature, optical fibers free sections that were not fixed to HPTRM were set at both ends of the optical fibers. The collected data can be corrected according to the system instructions.

The dial gauges were set above the round plate every 10 cm in the radial direction to measure the vertical deformation, as shown in Fig. 3

After 17 days' grass-growing, and water was filled in expansive soil of the model every day. The dial indicator data were recorded every day and optical fiber system

Fig. 1. The physical arrangement of model test

Table 2. The mechanical properties of HPTRM

Thickness mm	Mass Per Unit Area $g{\cdot}m^{-2}$	Tensile strength $kN{\cdot}m^{-1}$	Average elongation %
4.2	455	57 × 39	18.3 × 14.6

Table 3. Parameters of expansive soils

Density $g{\cdot}cm^{-3}$	Dry density $g{\cdot}cm^{-3}$	Water content %	Liquid Limit%	Plastic Limit%	Free expansion rate%	<5 μm clay content%
2.07	1.78	16.5	58.91	23.46	70	53.74

data were collected every four days. The applicability of the optical fiber was judged by comparing the results of the optical fiber system and the dial indicator.

3 Results and Discussion

The anchor position was defined coordinate origin, and horizontal coordinates of the dial indicator are form −50 cm to 50 cm. The vertical deformation of the dial indicator position can be directly measured by dial indicator (hereinafter the "measurements"). Strains of the optical fibers can be obtained with NXB-6050, while using the Pythagorean Theorem, vertical deformation of the specified position can be computed with

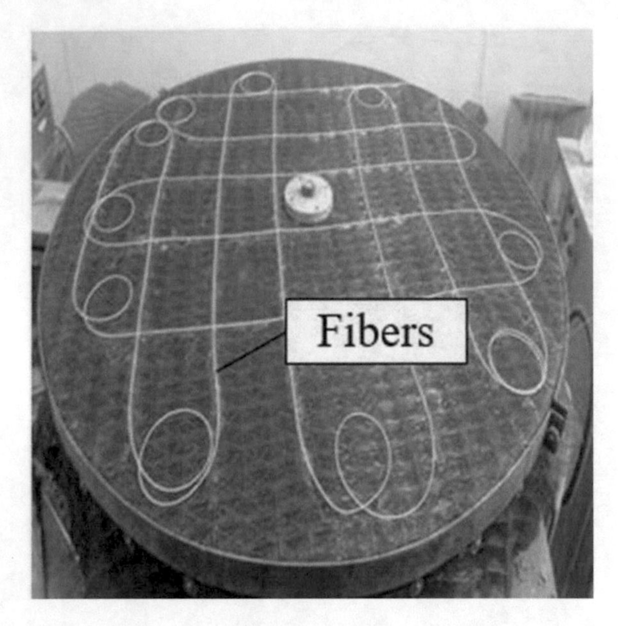

Fig. 2. Laboratory test fiber rollout

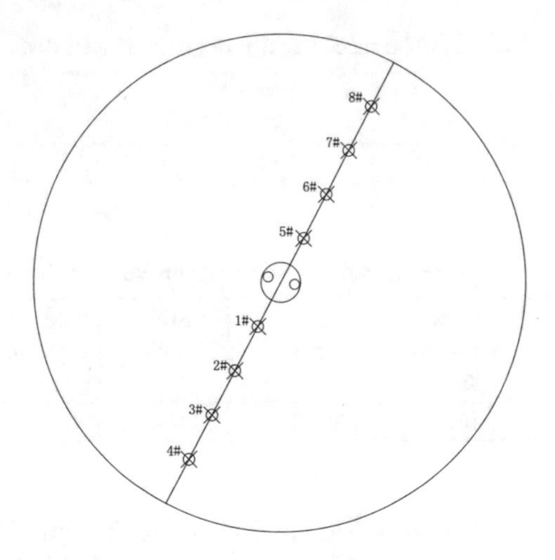

Fig. 3. Dial indicator layout

strains (hereinafter the "calculated values"). According to the calculation, the vertical displacement of the dial indicator position was gotten in the radial direction and compared with the measurements, as shown in Fig. 4.

Results show that there is no deformation at the position of anchorage and around the hoop in HPTRM model. What's more, the farther from the fixed support (such as anchorage and hoop), the greater deformation there will be. The calculated values and measurements are agreement, which the maximum difference between the two curves is

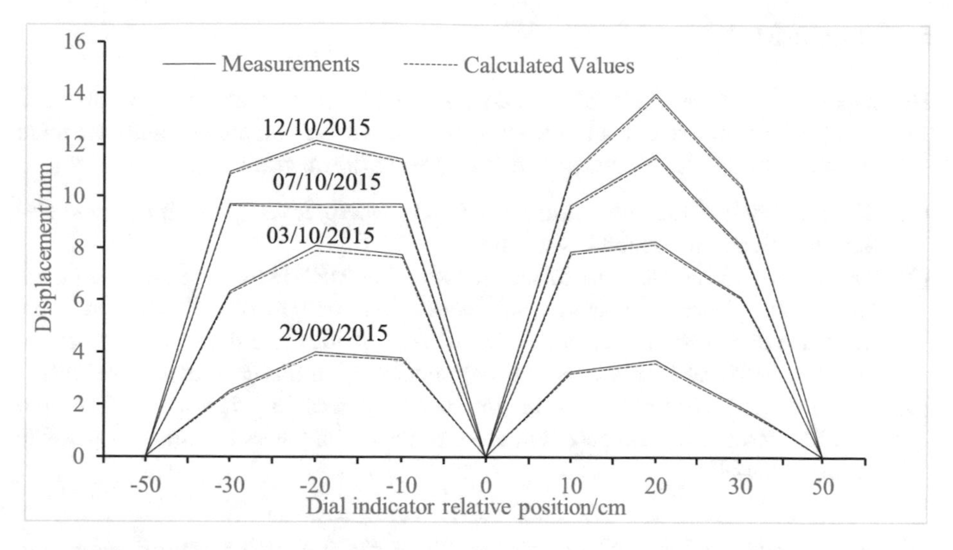

Fig. 4. Comparison of the vertical displacement of nodes along the radial direction

merely 0.12 mm in same position. The relative deviation is lower than 5% between calculated values and measurements. The relative deviation gradually decreases with the time increasing of filling water, and the minimum value is less than 1%. The reason for the relative deviation in the early stage is that the soil and HPTRM are not in close contact. With the continuous expansion of the soil during the test, the vertical deformation of soil and geonet gradually tends to synchronize.

4 Feasible Verification

Based on the monitoring slope protection data of ARVS system, Fan (Fan 2016) concluded several obvious advantages of the system. Field tests and the indoor tests mentioned above use the same soil samples and HPTRM. Dial indicator was placed every 1 m to measure the vertical deformation of expansive soil. And the vertical deformation is used to calculate the extension of the HPTRM. The maximum vertical deformation detected by the dial indicator is about 4 mm. And the calculated strains rate is 8×10^{-6}, which is far less than the ultimate elongation (Table 2). When the optical fibers and geonet are deform synchronously, the force of the optical fibers is about 0.1 N, far less than the ultimate tensile strength (Table 1). These results show that the distributed optical fiber sensors is applicable to deformation monitoring of expansive soils protected by HPTRM.

5 Conclusion

This paper analyzed the results of laboratory and field tests to study the applicable and practicality of distributed optical fiber sensors in expansive soil slope engineering. On the basis of analysis, the following conclusion were drawn from.

(1) The calculated values are consistent with the measurements, which demonstrates that the calculation method is reliable.
(2) The results of the site experiments showed that HPTRM can provide sufficient constraining force to expansive soil slope. The strength of HPTRM and fibers meets the deformation demand of the material. It illustrate the practicality of the distributed optical fiber sensors for deformation monitoring of expansive soils.
(3) The accuracy of the test can be improved by weaving the optical fiber into HPTRM and placing the geonet in close contact with the soil as much as possible during installation.

Acknowledgments. This research was supported by National Natural Science Foundation of China (NSFC) (No. 51369006).

References

Cliftion, D.H.: Full-scale testing of three innovative levee strengthening systems under overtopping condition. Master thesis, Jackson State University (2010)

Kang, C.Y.: Experimental study of erosion on expansive soil slope strengthened by High Performance Turf Reinforcement Mat (HPTRM) system in Nanning. Master thesis, Guangxi University (2015)

Xian, S.H.: Research on the protective effect of anchored reinforced vegetation system on expansive soil slope surface. Master thesis, Guangxi University (2016)

Li, H.Q., Sun, H.Y., Liu, Y.L., et al.: Application of optical fiber sensing technology to slope model test. Chin. J. Rock Mechan. Eng. (2008). https://doi.org/10.3321/j.issn:1000-6915.2008.08.022

Wang, B.J., Li, K., Shi, B., et al.: Simulation experiment for distributed fiber monitoring on deformation of soil slope. Chin. J. Rock Mechan. Eng. (2009). https://doi.org/10.3969/j.issn.1004-9665.2010.03.006

Zhu, H.H., Yin, J.H., Hong, C.Y., et al.: Fiber optic based monitoring technologies of slope engineering. Geotech. Investig. Surv. **3**, 6–10 (2010)

Fan, G.: Experimental study on protective effect of ARVS system on expansive soil slope. Master thesis, Guangxi University (2016)

Generalized Analysis of Under-Reamed Pile Subjected to Anisotropic Swelling Pressure

V. V. N. Prabhakara Rao(✉)

Department of Civil Engineering, V.R Siddhartha Engineering College, Kanuru, Vijayawada, A.P, India
vvnpjee@gmail.com

Abstract. Many pioneering foundation techniques have been suggested for plummeting the unfavorable heave of foundations placed in expansive clays. This paper presents another innovative technique of the determination of load creating a specified amount of settlement and load which maintains no uplift/no settlement. In the present study a slip layer of varying thickness is introduced along the pile periphery and depth of swelling zone, intensity of anisotropic swelling pressure are varied and stress based analysis is conducted and load creating a known amount of settlement and load which maintains no uplift/no settlement are found out. It is accomplished that (a) Increase in expansive zone increases, load required for known amounts of settlement increases for a given slip element thickness, bulb diameter and swelling pressure (b) The soil heave and pile uplift are immensely affected by the intensity of anisotropic swelling pressure intensity. (c) Provision of slip element below the pile tip reduces the pile upheaval.

Keywords: Foundation · Slip element · Heave · Swelling pressure Settlement

1 Introduction

It is well known that any volume change in Civil Engineering components is considered to be detrimental. This is more so with soil in view of its swelling properties. Expansive soil has been a subject of immense practical interest to Geotechnical Engineers. The obvious reason is that significant economic losses that often result from distress of buildings. Distress in the form of ground heave or pile uplift runs into risk of either failure of flooring or uplift of foundation system. According to a 1987 study, expansive soils in US inflict about a 9 billion in damages per year to buildings, loads, airports, pipelines and other facilities – more than twice the combined damages from the natural disasters such as earth quakes, floods, tornados and hurricanes (Coduto 1999). These expansive soils are widely spread and found in Argentina, Australia, South Africa, Burma, Canada, Cuba, Ethiopia, Israel, Spain, Ghana, Iran, Mexico, Rhodesia, Turkey, Venezuela, United States and India (Satyanarayana 1996). In India, Indian subcontinent, 20% of the land area is covered with expansive soils (Fig. 1a)

J. S. McCartney and L. R. Hoyos (Eds.): GeoMEast 2018, SUCI, pp. 165–187, 2019.
https://doi.org/10.1007/978-3-030-01914-3_14

popularly known as Black cotton soils and at some places, they are much known as "Regurs". These soils are formed due to the sub-aerial weathering of basalts in-situ and admixture of the weathered products with Iron and organic matter. Black cotton soils are predominantly found in the states of Madhya Pradesh, Gujarat, Akola region in Maharashtra (Fig. 1b), Andhra Pradesh, Tamil Nadu and border area of Uttar Pradesh adjoining Madhya Pradesh (Sinha 1996). Indian Black cotton soils are produced by the decomposition of basalts containing calcium and magnesium carbonates, potash, phosphates etc. In basic igneous rocks, black cotton soil is formed by decomposition of Feldspars and Pyroxene and in sedimentary rocks; it is a constituent of rock itself (Nayak 1985).

Aim of the Research is to investigate the effect of anisotropic swelling pressure on soil heave and pile uplift.

The Objective is to find out the loads required to (i) cause a given allowable/specified settlement of pile (Foundation), (ii) balance the soil expansion (i.e., no pile heave or settlement).

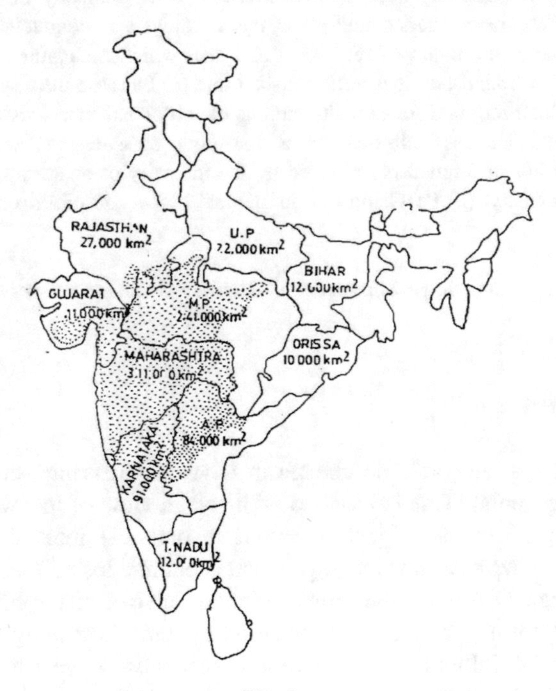

Fig. 1.

Fig. 1. (*continued*)

2 Literature Review

The available literature on soil expansivity deals with:
Anisotropic of soil expansion
Theoretical approaches
Physical models

2.1 Anisotropic of Soil Expansion

Anisotropy of soil Expansion received attention from researchers (Basma and Al-Akhras 1991; Gens and Alonso 1992; Sorochan and Losevan 1985; Viswanadham et al. 2009)

Experimental studies to establish anisotropy of soil expansion have been carried out in the past and the swelling anisotropy is found to have significant effect in the behavior of soil on a foundation (Basma and Al-Akhras 1991; Sorochan and Losevan 1985). The anisotropy is found to depend on stress history and contrary to general belief, activity and liquid limit are found to have no significant effect on swelling pressure (Basma and Al-Akhras 1991). A framework based on the distinction within the material of a micro-structural level where the basic swelling of the active minerals takes place and a macro-structural level responsible for major structural rearrangements and by coupling between the two levels, major features of the behavior like thermo-mechanical effects in concrete geostructures (piles, walls and slabs) can be reproduced (Gens and Alonso 1992). The soil resistance to horizontal deformation leads to increase in the contact pressure (Sorochan and Losevan 1985). The results of laboratory study performed on expansive soil reinforced with geofibers and demonstrated that discrete and randomly distributed geofibers are useful in restraining the swelling tendency of expansive soils and the mechanism by which discrete and randomly distributed fibers restrain swelling of expansive soil was explained with the help of soil–fiber interaction (Viswanadham et al. 2009).

2.2 Theoretical Approaches

Theoretical approach received attention from researches (Emilios et al. 2003; In-Mo Lee 1999; Poulos and Davis 1980).

Nonlinear numerical analysis establishes the load displacement relationship for various layouts of pile groups and the affect of spacing on ultimate bearing capacity, pile stiffness for both test and reaction piles, (Emilios et al. 2003).

Error back propagation neural network to predict the ultimate bearing capacity of piles based on simulated model pile load tests, the results of in situ pile load tests obtained from a literature survey were also used for training and testing alternatively and indicated that neural networks can be effectively used for pile capacity prediction, (In-Mo Lee 1999).

The Theoretical approach for analyzing piles installed in expansive soils using elastic theory noticed the soil swelling profile to vary almost linearly with depth from maximum at the surface to Zero at the base of clay and the enlarged base has greatest influence when the bulb is situated at or near the base of swelling soil, further, as diameter of shaft increased pile movement decreased (Poulos and Davis 1980).

2.3 Physical Modeling

Physical Modeling has received wide acceptance from the researches (Challa and Poulos 1991; Prakash et al. 1988; Chummar 1988; Fan et al. 2007; Lee and Park 2008; Katti 1979; Laloui et al. 2006; Phanikumar et al. 2004; Rama Rao and Smart 1977; Mohamedzein et al. 1999).

Swelling of the soil may induce substantial tensile force in the pile and the swelling soil profile varies almost linearly with depth from a maximum at the surface to zero at the base whereas the pile head movement increases as the soil heave increases (Challa and Poulos 1991). The Net uplift load and movement of the pile heads are over estimated by Poulos and Davis (Prakash et al. 1988). Under reamed piles are superior foundation when (a) Expansive soil is present in the zone of moisture variation and (b) The bearing capacity is less than the swell pressure of the soil (Chummar 1988). The elastic differential equations of load-transfer of single pile either with applied loads on pile-top or only under the soil swelling were established, respectively, based on the theory of pile-soil interaction and the shear-deformation method (Fan et al. 2007). New equivalent pile load–head settlement curve, considering the elastic shortening under top-down load which is similar to the pile load–head settlement curve is obtained by the static pile load test before the pile yields. The new method can be used to effectively estimate the pile head settlement as well as pile capacity by using the bi-directional load test results (Lee and Park 2008). The Swelling Pressure in black cotton soils is due to the presence of montmorillionite type clay mineral and it is possible to suppress the swelling & swelling pressure of black cotton clays by the addition of organic chemicals and CNS process is found to be effective in protection of earth retaining structures from high lateral pressure (Katti 1979). Coupled multi-physical finite element model is able to reproduce the most significant thermo-mechanical effects in concrete geostructures (Piles, walls and slabs) which is eventually an environmentally friendly way of cooling and heating buildings (Laloui et al. 2006). A granular pile anchor-foundation (GPAF) system arrests heave and improves the overall engineering behavior viz., reduces the heave of the expansive clay beds by about 96% and increases the un-drained strength by about 20% (Phanikumar et al. 2004). The Swell pressure measurements are subjected to systematic underestimate resulting from the flexibility of the components of the apparatus corresponding to approximately 1 percent volumetric strains and this error only be eliminated by a sophisticated and novel servo mechanism system (Rama Rao and Smart 1977). Parametric studies on two dimensional axi-symmetric finite element based model for analysis of a soil, nonlinear elastic,–pile, linearly elastic, system in expansive soils have been developed and predictions of the model are compared to the results of field experiments from two expansive soil sites (Mohamedzein et al. 1999).

2.4 Review of Theoretical Developments

From an over view of theoretical developments it can be seen that:
There are field trials to relate soil expansivity to pile uplift.
There are analytical methods to predict pile uplifts based on elastic theories.
There are construction techniques to limit foundation uplift.

Materials are developed which will neutralize the effect of swelling soils on foundations.

2.5 Limitations of These Methods

Generalizing the results of field trails may not hold good due to non homogeneity of soils.

Effect of anisotropic nature of swelling pressure (more closer to reality) was never attempted.

Effect of thickness of slip element had not been touched upon.

No studies had been made to visualize the case where depth of expansive soil is greater than length of pile.

2.6 Closing Remarks

Now the contrast between analytical model and theoretical model is presented in the Table 1.
Formulation of a model to take care of these limitations:

In-view of the above mentioned formulations which will obviously take care of the limitations stated above, a physical model is developed and shown as Fig. 2.

Table 1. Contrast between analytical model and physical model.

Analytical model	Physical model
Subjected to limitations like Type of soil, Condition favorable to swelling	Parameters can be easily changed and effect of the change can be visualized within short time
When ever an analytical model fails it consumes long time to formulate another	Even though one model fails, failure can be analyzed and model suited to given conditions can be reengineered quickly

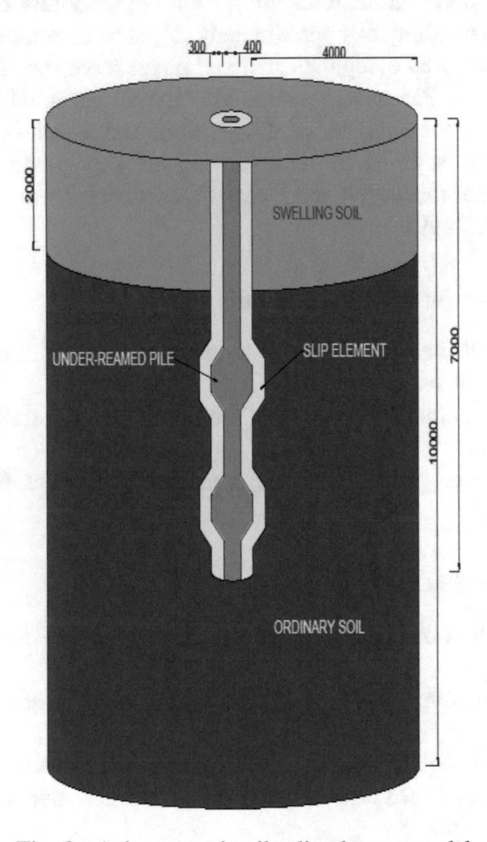

Fig. 2. Axi-symmetric pile slip element model

A parametric study is thus necessary to under stand the effect of anisotropic swelling pressure on behavior of plies placed in expansive soil. The study should ideally include the variables as (1) Diameter of Under-reamed pile, (2) Length of pile, (3) Effect of slip element & its thickness, (4) Depth of expansive soil (5) Intensity of Swelling Pressure including effect of anisotropic nature in swelling pressure.

3 Scope

3.1 Scope of the Present Work

The available literature establishes that there is adequate knowledge of physical modeling such as qualitative and semi quantitative methods. However analytical/ numerical approaches have limited attention from the researchers. Development of improved models is the need of the hour. Further, in order to stimulate the soil in the field as closely as possible, study on an-isotropy of soil with respect to swelling is necessary. It is therefore focused at analyzing the under-reamed piles installed in expansive soils subjected to an-isotropic swelling pressure in the following manner: Introduction of expansive soil, collection, assimilation, and classification of literature dealing with uplift of foundations, theoretical developments under "Literature Review", methodology, presentation of the results, analysis of the results and conclusions.

3.2 Problem Definition

The objective is to determine the variation of loads limiting the settlement to a range of generalized values as well as loads required to maintain the equilibrium (i.e. uplift/settlement). Therefore the need is to evaluate realistically, the loads on the doubly under-reamed pile coated with a slip element around its periphery due to anisotropic swelling of soil when it is embedded in soil of varying expansive soil layer. The diameter of the pile shaft is kept constant whereas the thickness of Slip layer, diameter of the under-reamed bulb and length are varied. In order to understand the effect of anisotropic swelling pressure on pile displacements, swelling pressure intensity was applied and a stress based analysis supported by FEM was performed and the corresponding load and settlement behavior is reported.

4 Methodology

It is obvious that with the existing theories, the soil expansion cannot be predicted reasonably. Innovative techniques are thus very much necessary to diagnose the problem, the present method aims at this.

A 7 m long pile with a slip element around its periphery is placed in an expansive soil. When the soil undergoes expansion, pile uplifts due to the effect of swelling pressure (Constant swelling pressure profile) in order to simulate this condition stated above axi-symmetric model has been developed Fig. 2. The Methodology for implementation such that soil swelling pressure is amenable to soil expansivity/pile (Foundation) uplift is to conduct a parametric study assisted by ANSYS. A cross section of the above-mentioned model is prepared for analysis and discretization of which leaves 198 soil elements, 18 pile elements and 18 slip elements. Now upon this model anisotropic swell pressure intensity in form of thermal stress is applied at the center of each expansive soil element, due to which obviously pile gets uplifted.

A parametric study is performed to understand the effect of anisotropic swelling pressure on the behavior of piles placed in expansive soil. The study should ideally have the variables as (a) Diameter of Under-reamed pile, (b) Length of pile, (c) Effect of slip element, (d) Depth of Expansive soil & (e) Intensity of anisotropic swelling pressure intensity.

Design constants obtained from laboratory tests are mentioned in the Table 2 and assumed range of variables are mentioned in the Table 3.

Table 2. Design Constants: Tested at geotechnical and concrete technology labs of VR Siddhartha Engineering College, Kanuru, Vijayawada.

Parameter	Value
Unit weight of ordinary soil	1.74 kN/m^3
Modulus of elasticity of ordinary soil	10000 kN/m^2
Unit weight of concrete (Pile elements)	2.5 kN/m^3
Modulus of elasticity of concrete	2.50E+07 kPa
Modulus of elasticity of expansive soil	5000 kN/m^2
Modulus of elasticity of slip element	1000 kN/m^2(Assumed),
Poisons ratio of soil and concrete	0.45
Thermal coefficient of expansion	0.002 (In the X-direction) and 0.001 (In the Y-direction).

Table 3. Ranges of Variables assumed:

Parameter	Range
Under-ream bulb Diameter	600 mm to 1000 mm
Length of pile	7000 mm
Slip element thickness	0 mm to 500 mm
Depth of swelling soil	2000 mm to 10000 mm
Intensity of anisotropic swelling pressure intensity	Swelling pressure in vertical direction = 0.1 kN/m^2, Swelling pressure in horizontal direction = 0.05 kN/m^2

4.1 FE Formation (Axi-Symmetry)

A double bulb Under–reamed pile of 7000 mm length (Double-bulb) is considered and placed in expansive soil which extends radially 4000 mm in all direction from the outside boundary of a slip element (Thickness of slip elements is varied between 0 mm to 500 mm). A hard stratum is assumed at 10 m below ground level. In the entire analysis, axial loads of different magnitudes are placed on the pile top. Thus, an axi-symmetric problem Fig. 2. is realized. The whole model when cut by any plane passing through the centroidal axis of the pile leads to discretization. Anisotropic swelling pressures of various intensities are applied at the centre of each expansive soil element.

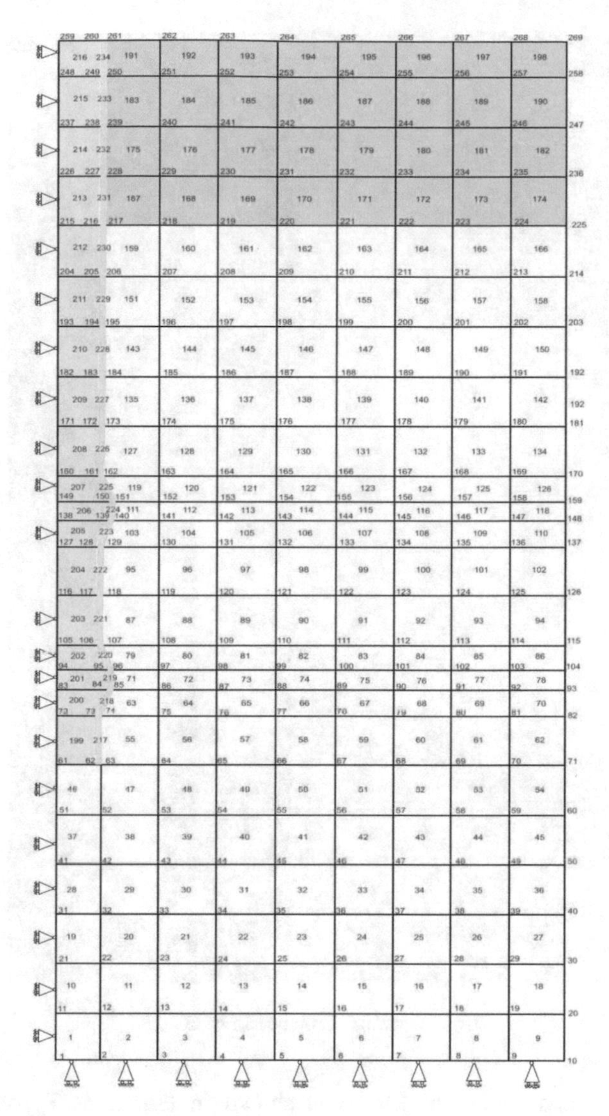

Fig. 3. Descretization of model

4.2 Discretization

Under–reamed Pile with a slip element around its periphery is placed in soil and is discretized into 198 soil elements, 18 pile elements and 18 slip elements totaling 234 quadrilateral elements. The geometry of the elements changes when ever dissimilar materials encounter (Overall iso-parametric property elements is maintained). All the elements in the discretization are straight sided elements. The discretization with different elements is presented in Fig. 3.

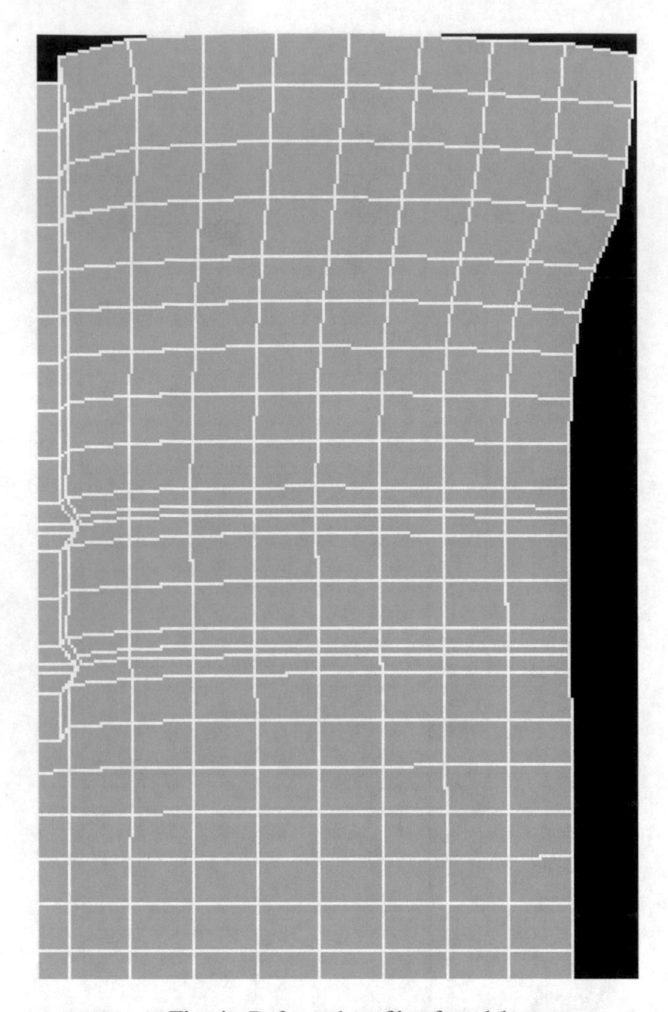

Fig. 4. Deformed profile of model

While, Deformed Profile of Model is shown in Fig. 4, & Typical nodal Displacements are shown in Fig. 5.

Design Criteria with appropriate rationale are mentioned in the Table 4.

Displacement Values at Node e259 of 7000mm Under-Reamed Pile (600mm) Embedded in 2000mm Deep Expansive Soil, Sp = 0.1N/mm2,0.05N/mm2, Thickness of Slip Element = 5mm

```
259 0.0000E+00-0.1509E-01
260 0.1412E-05-0.1499E-01
261 -0.5224E-03-0.1129E-01
262 0.1551E-02 0.7309E-02
263 0.5717E-02 0.1171E-01
264 0.1026E-01 0.1321E-01

***** POST1 NODAL DISPLACEMENT LISTING *****

LOAD STEP   1 ITERATION=   1 SECTION= 1
TIME=  0.00000E+00   LOAD CASE= 1

THE FOLLOWING X,Y,Z DISPLACEMENTS ARE IN NODAL
COORDINATES

NODE    UX      UY
265 0.1466E-01 0.1284E-01
266 0.1894E-01 0.1128E-01
267 0.2288E-01 0.8735E-02
268 0.2602E-01 0.5867E-02
269 0.2796E-01 0.3753E-02

MAXIMUMS
NODE    258     225
VALUE  0.2862E-01-0.1600E-01
```

Fig. 5. Typical nodal displacements

Table 4. Design Criteria and Rationale

SL. No	Design Criteria	Rationale
1	Maximum depth of swelling soil is 10 m, Width of soil continuum is equal to 4 m	In India the black cotton soils from transported origin are known to exist up to 8 m or more, Ranjan and Rao (2000). In view of this the depth of expansive soil is assumed as 10 m
		A number of trials were conducted to determine the limiting width of Soil continuum and the Q_0 values are influenced marginally only when the width of soil continuum is 4 m beyond the slip element
2	Length of the pile is 7 m	Choice is based on inspiration got from the model studies conducted by Sorochan and Losevan (1985)
4	Soil swelling pressure is anisotropic	To simulate a model as closely to natural condition as possible

(*continued*)

Table 4. (*continued*)

SL. No	Design Criteria	Rationale
5	Swelling pressure is analogous to temperature stress	In clays thermal conductivity is observed to increase with moisture content and there after the thermal connectivity reduces gradually as moisture content increases at 100% saturation Juneau et al. (1989)
		Instrument for in-situ measurement of thermal conductivity of soil is developed and the water movement in response to temperature gradient and the thermal properties of soil depend upon several factors such as soil constituents, moisture content etc., Sudhindra et al. (1988)
		With temperature rise and decrease of moisture content soil shrinks and with decrease in temperature and rise of water content soils swells. Thus it can be inferred that swelling pressure is analogous to temperature stress
6	Slip element is made of soft material.	Slip element, assumed to be soft material, is introduced all along pile periphery to induce shear deformation between soil & pile stress. The Proposed thin layer elements is a linear model of an interface element. To allow relative movement between soil & structure it requires a low shear modulus, so that shear deformation takes place
		But for linearly electric material, there is definite relation between Young's Modulus (E) & Shear Modulus (G). $E = 2G(1 + \mu)$ … (1) Hence $E \propto G$ … (2) Hence by reducing the value of 'E', we get reduced values 'G'.
		The property of this thin zone mainly depend upon the parent material properties. Hence for proposed thin lager elements we assumed $E_{SLIPELEMENT} = \left(\frac{1}{10}\right) * E_{PARENTMATERIAL}$ … (3)
7	Constant swelling pressure profile is assumed i.e. swelling pressure remains constant with depth.	Laboratory tests on expansive soils obtained from Akola, Maharashtra, India exhibited constant swelling pressure profile Chaudhari (1991).

(*continued*)

Table 4. (*continued*)

SL. No	Design Criteria	Rationale
8	A four noded iso-parametric soil element of 500 mm * 500 mm is used.	Four noded iso-parametric element is selected to allow linear variation of swelling across each element and to bring out accuracy in results The thickness of slip element varied from 0 to 500 mm. which means no slip condition when thickness of slip element is zero. Maximum thickness is slip elements is limited to 500 mm as width of soil element used in the analysis is 500 mm. Thus the ratio of slip element 'Ts' to width of soil element 'B' is varied between 0.0 and 1.0 to understand the optimum thickness of slip elements which result is minimum pile uplift
9	Settlement of the pile under working load is 15 mm.	Swelling Pressure was measured at every 0.5 m and a constant swelling pressure profile was reported by Chaudhari (1991). Thus with a view to simulate this condition, each element selected is of size 0.5 m * 0.5 m. Aspect ratio of soil elements is maintained as one, since the displacements vary at about same rate in each direction, the closer the aspect ratio to unity, the better the quality of solution
10	Axi-symmetric problem	The analysis deals with a double under-reamed pile, symmetric about longitudinal axis, placed in a expansive soil extending equally in all directions up to 4000 mm from the outside boundary of slip element attached to cylindrical shaped pile. The material remains the same in all directions irrespective of the angle subtended at the central the pile by any two given planes. Thus this is a 3D problem which can be best represented by axis-symmetric problem i.e. axi-symmetric solid is placed under axi-symmetric material. The intensity of stresses applied in vertical and horizontal direction of dissimilar. The same intensity of stress is applied on all soil elements. Therefore the problem can be best visualized as axis-symmetric material subjected to an- isotropic stress intensity

5 Results and Discussion

Results and Discussion are presented as under:

5.1 Results

Q_0 Values with the variation in pile lengths for (i) D_b = 600 mm; (ii) T_s = 300 mm and (iii) An-isotropic swelling pressure P_S intensity 0.1 kN/m^2 (Vertical), 0.05 kN/m^2 (Horizontal) are presented in the Table 5.

Table 5. Q_0 Values for Different Pile Lengths for $D_b = 600$ mm, $T_s = 300$ mm and $Sp = 0.1$ kN/m^2,0.05 kN/m^2

Depth of expansive soil in m	Length of Pile = 5 m	Length of Pile = 7 m	Length of Pile = 8 m
2	31.42	6.28	0
4	138.23	62.83	54.25
6	245.04	226.18	213.63
8	420	414.69	351.86
10	534	502.65	376.99

Discussion on Table 5.
For a given pile length Q_0 increases with the increase in the depth of swelling zone. As the pile length increases the load required to maintain equilibrium decreases.

Q_{15} Values with the variation in pile lengths for (i) $D_b = 600$ mm; (ii) $T_s = 300$ mm and (iii) An-isotropic swelling pressure P_S intensity 0.1 kN/m^2 (Vertical), 0.05 kN/m^2 (Horizontal) are presented in the Table 6.

Table 6. Q_{15} Values for Different Pile Lengths for $D_b = 600$ mm, $T_s = 300$ mm and $Sp = 0.1$ kN/m^2,0.05 kN/m^2

Depth of Expansive Soil in m	Length of Pile = 5 m	Length of Pile = 7 m	Length of Pile = 8 m
2	238.78	270.18	299.04
4	307.87	320.44	389.56
6	452.4	464.96	490.09
8	565.49	615.75	628.32
10	647.17	697.43	785.4

Discussion on Table 6.
For a given pile length Q_{15} increases with the increase in the depth of swelling zone As the pile length increases the load required to maintain equilibrium increases

Effect of Lateral extent of soil continuum on Q_0 values for $T_s = 300$ mm & $Sp = 0.1$ kN/m^2, 0.05 kN/m^2 is presented in the Table 7.

Table 7. Effect of Lateral extent of soil continuum on Q_0 values for $T_s = 300$ mm and $Sp = 0.1$ kN/m^2, 0.05 kN/m^2

Depth of Expansive Soil Z_e in m	Under-ream bulb diameter in mm	Lateral extent of soil continuum Equal to 3 m	Lateral extent of soil continuum Equal to 4 m	Lateral extent of soil continuum Equal to 5 m	Lateral extent of soil continuum Equal to 6 m
2	600	7.66	6.28	5.93	5.91
2	800	11.49	9.42	8.9	8.87
2	1000	15.32	12.56	11.87	11.83

(continued)

Table 7. (*continued*)

Depth of Expansive Soil Z_e in m	Under-ream bulb diameter in mm	Lateral extent of soil continuum Equal to 3 m	Lateral extent of soil continuum Equal to 4 m	Lateral extent of soil continuum Equal to 5 m	Lateral extent of soil continuum Equal to 6 m
4	600	76.65	62.83	59.37	59.19
4	800	86.23	70.68	66.79	66.58
4	1000	91.99	75.4	71.25	71.03
6	600	275.95	226.19	213.75	213.52
6	800	291.29	238.76	225.62	225.39
6	1000	306.62	251.33	237.5	237.25
8	600	497.63	414.69	391.88	389.47
8	800	588.11	490.09	463.14	461.66
8	1000	663.5	552.92	522.5	520.3
10	600	603.18	502.65	488	475
10	800	708.74	590.62	578.13	558.24
10	1000	776.6	647.17	627.57	611.51

Discussion on Table 7.

The Q_0 values increase when Lateral extent of soil continuum is less than 4 m where as it decreases as the Lateral extent of soil continuum increases beyond 4 m even though the decrease is marginal.

The Load Q vs settlement ρ for a given (i) An-isotropic swelling pressure P_S intensity 0.1 kN/m^2 (Vertical), 0.05 kN/m^2 (Horizontal), (ii) 0 mm thick slip layer (T_s) and with the variation of (i) Depth of swelling soil Z_e and (ii) Under-reamed bulb diameter D_b is depicted in Fig. 6.

Discussion on Fig. 6.

Load Q required for no heave/no settlement is the least for axi-symmetric pile model with no slip element around pile periphery.

Load Q required for 15 mm Settlement is the maximum for axi-symmetric pile model with no slip element around pile periphery.

As the depth of swelling soil increases load required at the pile top increases.

As the bulb diameter increases load required at the pile top increases.

When the swelling soil depth increases from 2 m to 10 m the load at the pile top for balancing soil heave increases by 12 times.

When the swelling soil depth increases from 2 m to 10 m the load at the pile top for creating 15 mm settlement increases by 2 times.

The Load Q vs settlement ρ for a given (i) An-isotropic swelling pressure P_S intensity 0.1 kN/m^2 (Vertical), 0.05 kN/m^2 (Horizontal), (ii) 5 mm thick slip layer (T_s) and with the variation of (i) Depth of swelling soil Z_e and (ii) Under-reamed bulb diameter D_b is depicted in Fig. 7.

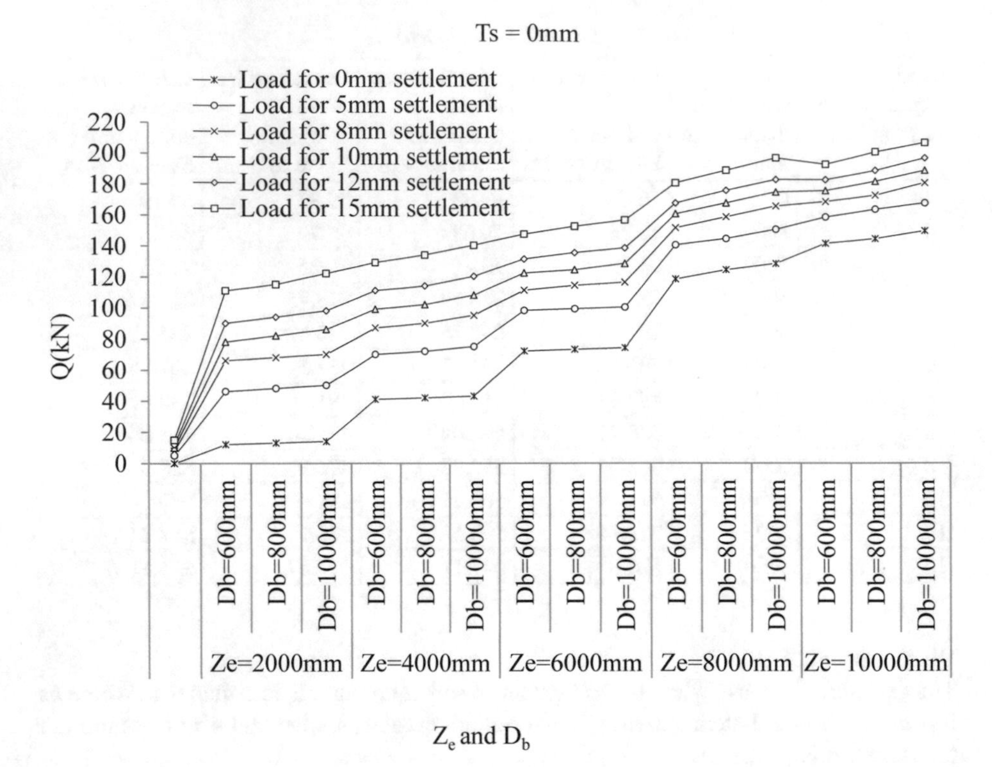

Fig. 6. Load Q vs settlement ρ for an-isotropic swelling pressure P_S intensity 0.1 kN/m^2 (Vertical), 0.05 kN/m^2 (Horizontal) and 0 mm thick slip layer (T_s)

Discussion on Fig. 7.

Load Q required for no heave/no settlement is the least for axi-symmetric pile model with 5 mm slip element around pile periphery.

Load Q required for 15 mm Settlement is the maximum for axi-symmetric pile model with 5 mm slip element around pile periphery.

As the depth of swelling soil increases load required at the pile top increases.

As the bulb diameter increases load required at the pile top increases.

When the swelling soil depth increases from 2 m to 10 m the load at the pile top for balancing soil heave increases by 19.2 times.

When the swelling soil depth increases from 2 m to 10 m the load at the pile top for creating 15 mm settlement increases by 1.84 times.

The Load Q vs settlement ρ for a given (i) An-isotropic swelling pressure P_S intensity 0.1 kN/m^2 (Vertical), 0.05 kN/m^2 (Horizontal), (ii) 50 mm thick slip layer (T_s) and with the variation of (i) Depth of swelling soil Z_e and (ii) Under-reamed bulb diameter D_b is depicted in Fig. 8.

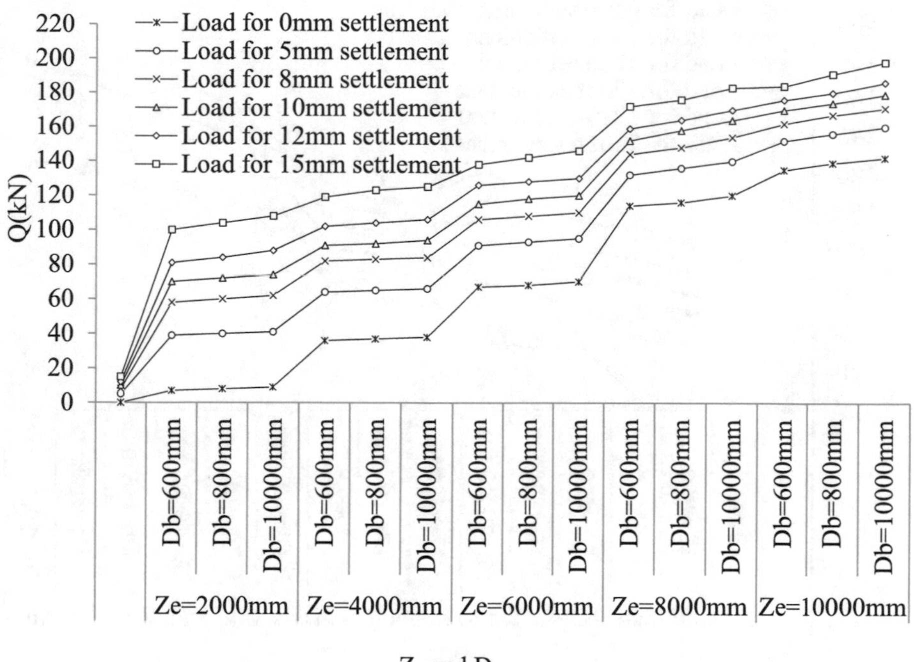

Fig. 7. Load Q vs settlement ρ for an-isotropic swelling pressure P_S intensity 0.1 kN/m^2 (Vertical), 0.05 kN/m^2 (Horizontal) and 5 mm thick slip layer (T_s)

Discussion on Fig. 8.

Load Q required for no heave/no settlement is the least for axi-symmetric pile model with 50 mm slip element around pile periphery.

Load Q required for 15 mm Settlement is the maximum for axi-symmetric pile model with 50 mm slip element around pile periphery.

As the depth of swelling soil increases load required at the pile top increases.

As the bulb diameter increases load required at the pile top increases.

When the swelling soil depth increases from 2 m to 10 m the load at the pile top for balancing soil heave increases by 19.2 times.

When the swelling soil depth increases from 2 m to 10 m the load at the pile top for creating 15 mm settlement increases by 2.12 times.

The Load Q vs settlement ρ for a given (i) An-isotropic swelling pressure P_S intensity 0.1 kN/m^2 (Vertical), 0.05 kN/m^2 (Horizontal), (ii) 100 mm thick slip layer (T_S) and with the variation of (i) Depth of swelling soil Z_e and (ii) Under-reamed bulb diameter D_b is depicted in Fig. 9.

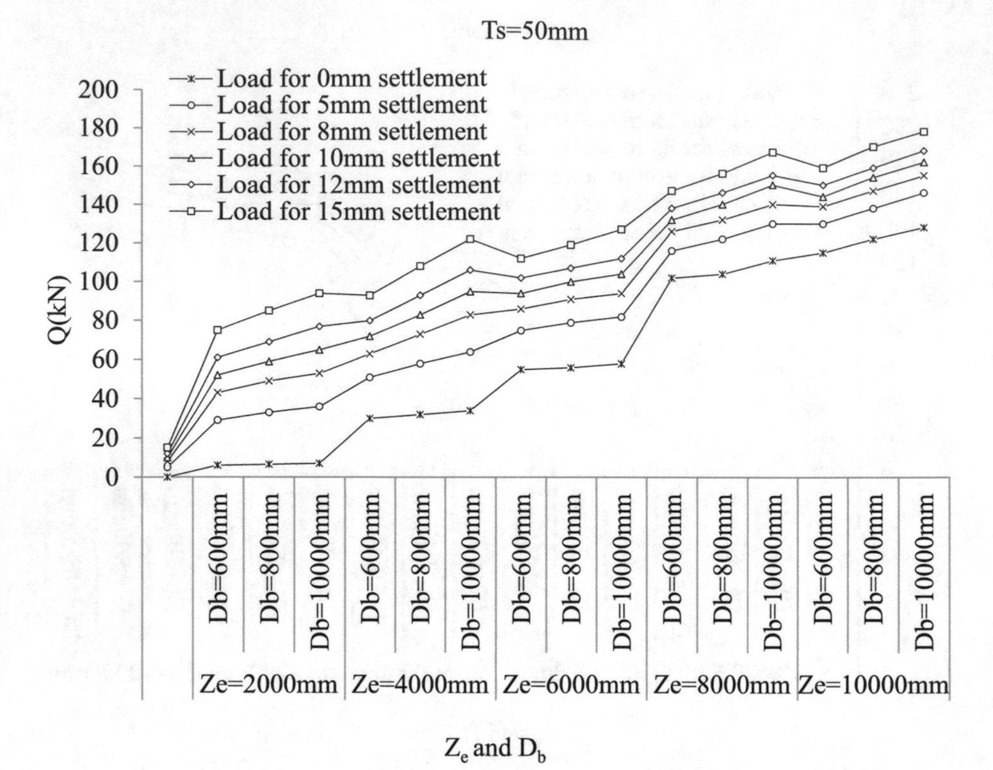

Fig. 8. Load Q vs settlement ρ for an-isotropic swelling pressure P_S intensity 0.1 kN/m^2 (Vertical), 0.05 kN/m^2 (Horizontal) and 50 mm thick slip layer (T_s)

Discussion on Fig. 9.

Load Q required for no heave/no settlement is the least for axi-symmetric pile model with 100 mm slip element around pile periphery.

Load Q required for 15 mm Settlement is the maximum for axi-symmetric pile model with 100 mm slip element around pile periphery.

As the depth of swelling soil increases load required at the pile top increases.

As the bulb diameter increases load required at the pile top increases.

When the swelling soil depth increases from 2 m to 10 m the load at the pile top for balancing soil heave increases by 23.11 times.

When the swelling soil depth increases from 2 m to 10 m the load at the pile top for creating 15 mm settlement increases by 2.29 times.

The Load Q vs settlement ρ for a given (i) An-isotropic swelling pressure P_S intensity 0.1 kN/m^2 (Vertical), 0.05 kN/m^2 (Horizontal), (ii) 300 mm thick slip layer (T_s) and with the variation of (i) Depth of swelling soil Z_e and (ii) Under-reamed bulb diameter D_b is depicted in Fig. 10.

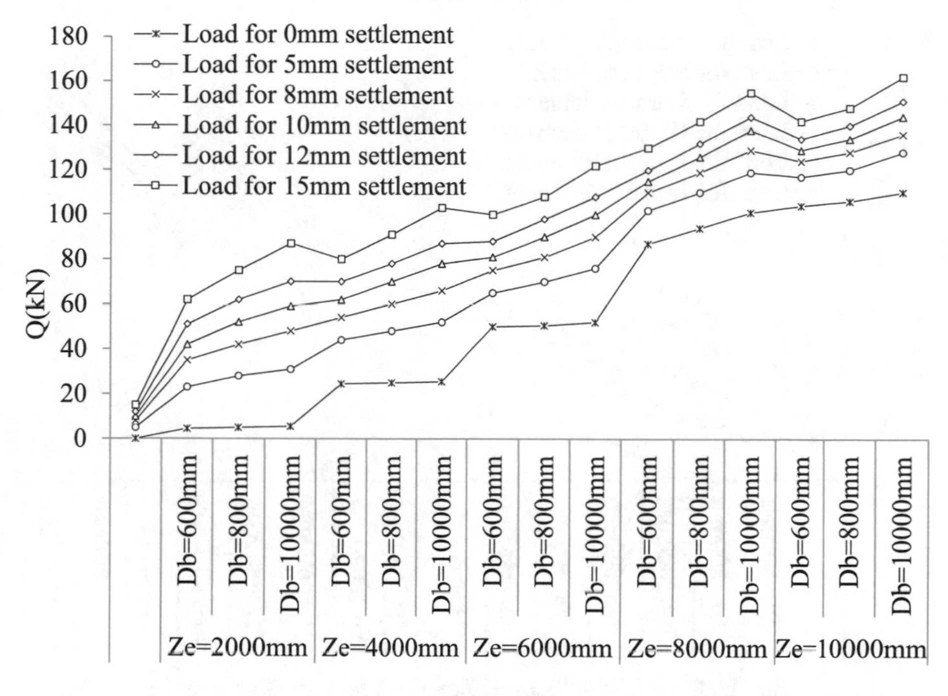

Fig. 9. Load Q vs settlement ρ for an-isotropic swelling pressure P_S intensity 0.1 kN/m^2 (Vertical), 0.05 kN/m^2 (Horizontal) and 100 mm thick slip layer (T_s)

Discussion on Fig. 10.

Load Q required for no heave/no settlement is the least for axi-symmetric pile model with 300 mm slip element around pile periphery.

Load Q required for 15 mm Settlement is the maximum for axi-symmetric pile model with 300 mm slip element around pile periphery.

As the depth of swelling soil increases load required at the pile top increases.

As the bulb diameter increases load required at the pile top increases.

When the swelling soil depth increases from 2 m to 10 m the load at the pile top for balancing soil heave increases by 160 times.

When the swelling soil depth increases from 2 m to 10 m the load at the pile top for creating 15 mm settlement increases by 2.56 times.

The Load Q vs settlement ρ for a given (i) An-isotropic swelling pressure P_S intensity 0.1 kN/m^2 (Vertical), 0.05 kN/m^2 (Horizontal), (ii) 500 mm thick slip layer (T_s) and with the variation of i) Depth of swelling soil Z_e and (ii) Under-reamed bulb diameter D_b is depicted in Fig. 11.

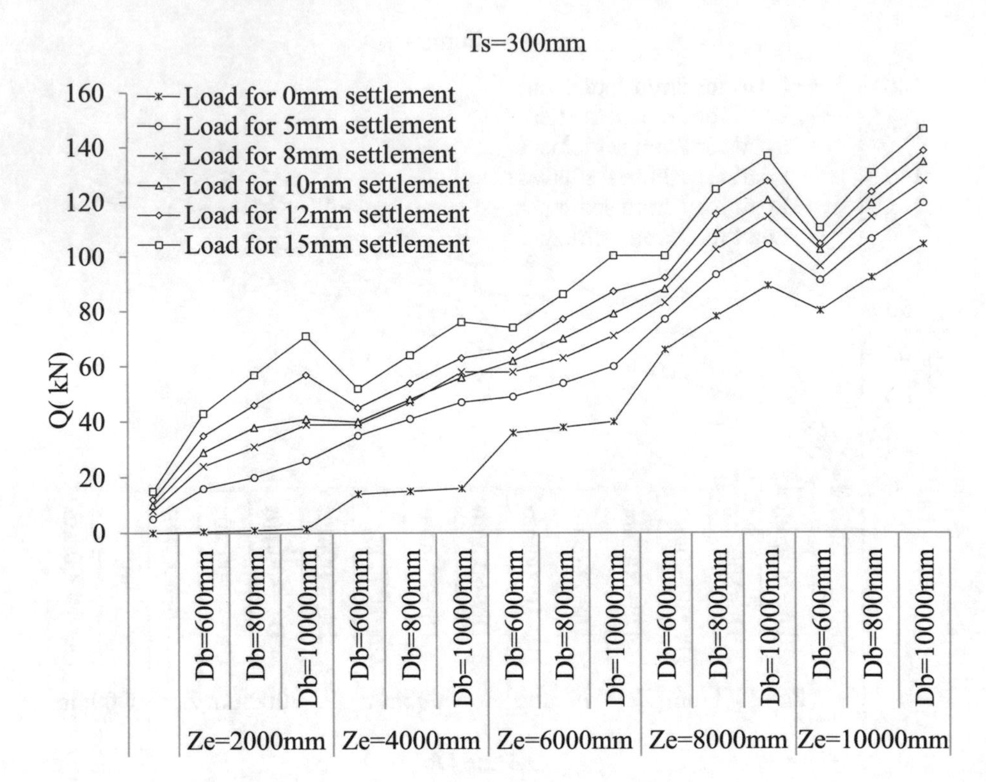

Fig. 10. Load Q vs settlement ρ for an-isotropic swelling pressure P_S intensity 0.1 kN/m^2 (Vertical), 0.05 kN/m^2 (Horizontal) and 300 mm thick slip layer (T_s)

Discussion on Fig. 11.

Load Q required for no heave/no settlement is the least for axi-symmetric pile model with 500 mm slip element around pile periphery.

Load Q required for 15 mm Settlement is the maximum for axi-symmetric pile model with 500 mm slip element around pile periphery.

As the depth of swelling soil increases load required at the pile top increases.

As the bulb diameter increases load required at the pile top increases.

When the swelling soil depth increases from 2 m to 10 m the load at the pile top for balancing soil heave increases from 0kN for all under-reamed pile diameters to 71 kN, 84 kN and 100 kN for 600 mm, 800 mm, 1000 mm respectively.

When the swelling soil depth increases from 2 m to 10 m the load at the pile top for creating 15 mm settlement increases by 2.56 times.

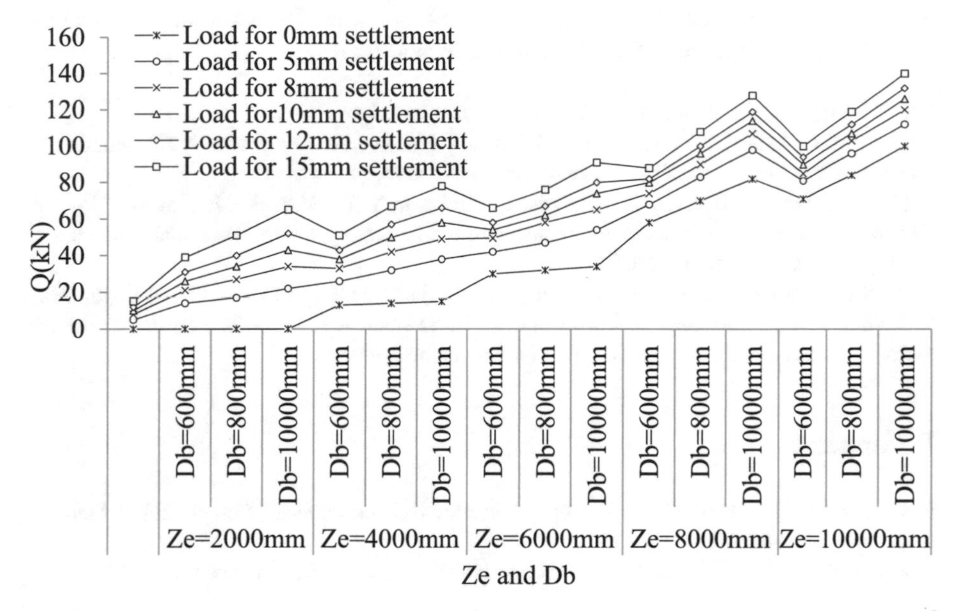

Fig. 11. Load Q vs settlement ρ for an-isotropic swelling pressure P_S intensity 0.1 kN/m^2 (Vertical), 0.05 kN/m^2 (Horizontal) and 500 mm thick slip layer (T_s)

6 Conclusions

The conclusions that are drawn from the present study may be summarized as follows:

The loads required to create a specified amount of settlement are observed to be more than loads required to maintain no heave condition at any swelling pressure intensity, for any slip element thickness, for any under reamed bulb diameter and for any given depth of swelling zone.

The introduction of slip element below the pile tip reduced the (a) Load creating a specified amount of settlement and (b) Load maintaining equilibrium by 25% in general.

As pile length increases the load required to maintain equilibrium decreases for a given thickness of swelling zone where as it increases when the depth of swelling zone increases.

As pile length increases the load required to create a specified amount of settlement increases for a given thickness of swelling zone where as it increases when the depth of swelling zone increases.

Q_O, Load required for no heave/no settlement is the highest for axi-symmetric pile model with no slip element around pile periphery and least for 500 mm slip element around pile periphery for 2000 mm deep swelling soil.

Q_{15}, Load required for 15 mm Settlement decreases by 2.85 times when the slip element thickness changes from 0 mm to 500 mm when the swelling soil depth is 2000 mm and under-reamed bulb diameter is 600 mm.

Q_{15}, Load required for 15 mm Settlement decreases by 1.99 times when the slip element thickness changes from 0 mm to 500 mm when the swelling soil depth is 10000 mm and under-reamed bulb diameter is 600 mm.

Acknowledgements. Author want to sincerely acknowledge:

(1) Shri. Chaudary G.B., (M.E.Geo-Technical), Rtd Assistant Professor, VJTI Mumbai for sharing information on swelling pressure profile at Akola, Maharashtra.

(2) Shri. Haroon Ali Khan, (M.S) Assistant Professor, V.R. Siddhartha Engineering College, Kanuru, Vijayawada, A.P, India for performing the laboratory test on soils and rendering all helping in preparing the manuscript.

(3) Shri. Jaya Krishna Sairam, (M.Tech. Structures) alumni, Department of Civil Engineering, V.R Siddhartha Engineering College, Kanuru, Vijayawada, A.P, India for performing the laboratory test on concrete and preparing the axi-symetric model.

References

Basma, A.A., Al-Akhras, N.: Anisotropy in swelling characteristics of clays. Indian Geotech. J. **25**(4), 429–444 (1991)

Challa, P.K., Poulos, H.G.: Behavior of single pile in expansive clay. Geotech. Eng. **22**(2), 189–216 (1991)

Prakash, C., Wardle, I.F., Chandrra, R., Balodhi, G.R., Price, G.: Performance of single under-reamed pile foundation supporting a single story structure in expansive soil. Indian Geotech. J. **18**(4), 340–355 (1988)

Chaudhari, G.B.: An enquiry report submitted to Chief Engineer, Public Works Department, Akola, Maharastra, India for the site of tehsil building at Daryapur. vol. 1, No. 2, p. 23 (1991)

Chummar, A.V.: For which soil conditions are under-reamed pile foundations superior to other foundation systems?. In: Indian Geotechnical Conference, pp. 121–128 (1988)

Coduto, D.P.:. Geotechnical Engineering Principals and Practice. Prentice Hall, pp. 655–664, 666–669, 671–672, 684–695 (1999)

Emilios, M., Christos, C., Poulos, T.A., Michael, K.G.: Numerical assessment of axial pile group response based on load test. Comput. Geotech. **30**(6), 505–515 (2003)

Fan, Z.-H., Wang, Y.-H., Xiao, H.-B., Zhang, C.-S.: Analytical method of load-transfer of single pile under expansive soil swelling. J. Central South Univ. Technol. **14**(4), 575–579 (2007)

Gens, A., Alonso, E.E.: A framework for the behavior of unsaturated expansive clays. Can. Geotech. J. **26**(6), 1013–1032 (1992). https://doi.org/10.1139/t92-120

Ranjan, G., Rao, A.S.R.: Basic and applied soil mechanics, 2nd edn, p. 702. New Age International Publishers, New Delhi (2000). Ch. 20

Lee, I.-M., Lee, J.-H.: Prediction of pile bearing capacity using artificial neural networks. Comput. Geotech. **18**(3), 189–200 (1996)

Lee, J.-S., Park, Y.-H.: Equivalent pile load–head settlement curve using a bi-directional pile load test. Comput. Geotech. **35**(2), 124–133 (2008)

Junejan, P.L., Gulhati, S.K., Singh, A., Venkatappa Rao, G.: An approach for determination of thermal conductivity of soils. Indian Geotech. J. **14**(4), 250 (1984)

Katti, R.K.: Search for solutions to problems in black cotton soils. In: Indian Geotechnical Journal, 20th Annual General Session, 1st I.G.S Annual Lecture, vol. 9, No. 1, pp. 76–78 (1979)

Laloui, L., Nuth, M., Vulliet, L.: Experimental and numerical investigations of the behavior of a heat exchanger pile. Int. J. Numer. Anal. Methods Geomech. **30**(8), 763–781 (2006). https://doi.org/10.1002/nag.499

Nayak, N.V.: Foundation Design Manual, 1st edn, p. 324. Dhanpat Rai Sons, Delhi (1985)

Poulos, H.G., Davis, E.H.: Pile Foundation Analysis & Design, Ch. 12, pp. 297, 299–301. John Wiley & Sons, New York (1980)

Phanikumar, B., Sharma, R., Srirama, R.A., Madhav, M.: Granular pile anchor foundation (gpaf) system for improving the engineering behavior of expansive clay beds. Geotech. Test. J. **27** (3), 279–287 (2004). https://doi.org/10.1520/GTJ11387. ISSN 0149-6115

Rama Rao, R., Smart, P.: Errors in swell pressure measurement. Proceedings of the First National Symposium on Expansive Soils, Session 1, pp. 4-1, 4-5 (1977)

Satyanarayana, B.: Swelling behavior and identification of expansive soils. In: National Seminar on Partially Saturated Soils and Expansive Soils, Kakinada, Andhra Pradesh, India, pp. 41, 43, 47, 48 (1996)

Sinha, U.N.: Quasi-over consolidation effect of expansive clays. In: National Seminar on Partially Saturated Soils and Expansive Soils, Kakinada, Andhra Pradesh, India, p. 49 (1996)

Sorochan, E.A., Losevan, A.N.: Pile behavior in swelling soils under horizontal load action. In: Plenum Publishing Corporation, New york, (Translated from Osnovania, Fundamenty I Mekhanika Gruntov, No. 3, PP. 10–12), Soil Mechanics & Foundation Engineering, pp. 89–93 (1985)

Sudhindra, C., Sharma, V.M., Dhavan, A.K., Gangadhar, R.B., Singh, P.P.: An instrument for the measurement of thermal conductivity of soil. Indian Geotech. J. **8**(2), pp. 186, 192, 193 (1988)

Tomlinson, M.J.: Foundation Design and Construction, 5th edn., pp. 423, 426. Longman, ELBS. Singapore (1986)

Viswanadham, B.V.S., Phanikumar, B.R., Mukherjee, R.V.: Swelling behavior of a geofiber-reinforced expansive soil. Geotext. Geomembr. **27**(1), 73–76 (2009). https://doi.org/10.1016/j.geotexmem.2008.06.002

Mohamedzein, Y.E.-A., Mohamed, M.G., El Sharief, A.M.: Finite element analysis of short piles in expansive soils. Comput. Geotech. **24**(3), 231–243 (1999). https://doi.org/10.1016/S0266-352X(99)00008-7

Author Index

J. S. McCartney and L. R. Hoyos (Eds.): GeoMEast 2018, SUCI, p. 189, 2019.
https://doi.org/10.1007/978-3-030-01914-3

Printed in the United States
By Bookmasters